C000212502

THE IMMORTAL YEW

THE IMMORTAL YEW

Tony Hall

Kew Publishing
Royal Botanic Gardens, Kew

First published in 2018 by the Royal Botanic Gardens, Kew, Richmond, Surrey, TW9 3AB, UK. www.kew.org

ISBN 978-1-84246-658-2

Distributed on behalf of the Royal Botanic Gardens, Kew in North America by the University of Chicago Press, 1427 East 60th St, Chicago, IL 60637, USA.

British Library Cataloguing in Publication Data
A catalogue record for this book is available from the British Library.

Production Management: Georgie Hills
Design and page layout: Christine Beard
Copy-editing: Jean Postle
Proofreading: Sharon Whitehead

Front cover photo: St Edward's Church, Stow-on-the-Wold, Gloucestershire
Back cover photos: top, Much Marcle Yew, Herefordshire; bottom, Martindale Yew, Cumbria
Photo opposite p 1: St Peter & St Paul Church, Cudham, Kent.

Printed in Italy by Printer Trento srl

For information or to purchase all Kew titles please visit shop.kew.org/kewbooksonline or email publishing@kew.org

Kew's mission is to be the global resource in plant and fungal knowledge, and the world's leading botanic garden.

Kew receives about one third of its running costs from Government through the Department for Environment, Food and Rural Affairs (Defra). All other funding needed to support Kew's vital work comes from members, foundations, donors and commercial activities, including book sales.

CONTENTS

Acknowledgements

I would like to thank the following for their help and support.

My partner Emma Crawforth, who visited many of the yews with me and read through my first drafts. Kew's publishing team, whose positivity encouraged me throughout this project.

Jake Davies-Robertson for turning the wonderful yew bowl illustrated in this book.

The following people and organisations, who allowed me to use their written work, property and gardens.

Alison Goulbourne, founder of the Gwenfrewi Project, which came to the rescue of a rundown Welsh church, giving it a second life, and preserving the ancient yews in the old churchyard. Edward Storey, a local poet, who found inspiration in the Discoed Yews. Earl Bathurst, custodian of the world's tallest yew hedge. Levens Hall, Kendal, with its inspirational topiary.

The National Trust, English Heritage and Historic Royal Palaces (including Hampton Court Palace), who preserve many of our heritage landscapes, which in turn protect many ancient trees.

The National Trust properties of Crom and Florence Court in County Fermanagh. And in the Republic of Ireland, the President of St Patrick's College, Maynooth.

Finally, all those involved in the churches and churchyards that gave their whole-hearted support to this project, and that cherish their ancient yews. These trees make up the largest percentage of the yews featured in this book.

Timeline of the European yew

It's hard to be precise about the history of the yew, but here are some key dates:

- **140 million years ago**
  Taxaceae (yew family) fossils formed.
- **15 million years ago**
  Taxus baccata (European yew) fossils formed.
- **200 BC**
  Herbalist Nikander describes the painful death caused by yew toxin.
- **53 BC**
  Catuvolcus, King of the Celtic Eburones tribe, poisoned himself with yew rather than become a Roman slave.
- **77 AD**
  First-known European topiary in Roman gardens, according to the Roman author Pliny the Elder.
- **1066**
  Battle of Hastings – King Harold killed by an arrow that supposedly pierced his eye, fired from a Norman yew longbow.
- **1215**
  Magna Carta signed, probably under the Ankerwycke yew (p 46).
- **1415**
  Battle of Agincourt – English archers with yew longbows proved to be decisive.
- **1536–41**
  Dissolution of the Monasteries, when Henry VIII ordered the destruction of monasteries, priories and convents in England, Wales and Ireland.
- **1692**
  Yew topiary garden planted at Levens Hall (p 136) and clipped annually to the present day.
- **1700s**
  Bathurst Estate Yew Hedge (p 52) planted – today it's probably the tallest yew hedge in the world.
- **1780**
  The original seedlings of Irish yew, from which all of today's Irish yews are descended, found near Florence Court (p 102).
- **1803**
  William Wordsworth wrote his famous poem *Yew Trees* about the 'Fraternal Four' yews of Borrowdale, Cumbria.
- **1897**
  The Yew-Trees of Great Britain and Ireland by John Lowe published – a classic guide to ancient yews.

- **1911**
  400,000-year-old yew spearhead, one of the world's oldest wooden artefacts, found at Clacton-on-Sea, Essex.
- **1960s**
  The world's oldest maze at Hampton Court Palace (p 116) re-planted with yews.
- **1964**
  Cancer drug paclitaxel discovered in Pacific yew bark samples by two Americans, Monroe Wall and Mansukh Wani.
- **1982**
  Mary Rose, Henry VIII's warship, which sank in battle in 1545, raised from the seabed. It was found to contain 29,000 artefacts, including more than 130 yew bows and staves.
- **1986**
  Anti-cancer drug Docetaxel, extracted from the leaves of the European yew, was patented and later approved for medical use.
- **1991**
  A yew longbow thought to be around 5,000 years old, found in the possession of the 'Ice Man', Ötzi, in the Tyrolean Alps.
- **1994**
  Synthetic cancer drug Taxol developed.
- **2000**
  Scientists at the Royal Botanic Gardens, Kew developed a method to diagnose yew poisoning by detecting the toxic taxane alkaloids in stomach contents. They used an analytical chemistry technique involving physical separation and mass analysis.
- **2002**
  A list of 50 Great British Trees compiled by The Tree Council to celebrate the Queen's Golden Jubilee. It includes 11 outstanding yews, several of which feature in this book, including the Ashbrittle yew (p 48), Llangernyw yew (p 144), Fortingall yew (p 104) and Borrowdale yews (p 60).
- **2014**
  Planting of the Yew Conservation Hedge at the Royal Botanic Garden Edinburgh begun – it is made from cuttings taken from over 2,000 European yews collected from threatened wild populations around the world, and from ancient and famous heritage trees.
- **2018**
  New yew trees are planted in churchyards for the Trees for Sacred Spaces campaign run by The Conservation Foundation with the aim of making London greener, healthier and more biodiverse.

INTRODUCTION

This book has taken me around Britain and Ireland, looking at a species of tree that is truly remarkable: the common or European yew, *Taxus baccata*. Some individual yew trees are among the oldest, if not the oldest, living things in Europe.

It was on a short break in Devon that I chanced upon a huge male yew tree in a churchyard in Kenn (p 128). I was stunned by its immense size. I knew the ancient Ankerwycke yew (p 46) from having visited it many times and wondered just how many other trees there were like these. Surprisingly the answer is quite a few.

With so many great old yew trees around the country, my plan changed and developed to visiting more of them than I had initially intended, including many quietly growing in wonderful rural, picturesque locations.

Originally, I was only planning to visit yews that had a circumference of 9 m (30 ft) or over. But during my research about these amazing arboreal giants, I found that there were many that were smaller in girth but equally interesting in different ways, with strange and unnatural shapes, each one unique, and with interesting stories attached to them.

So, it was difficult to decide which of the many trees I had visited should feature in this book. I have tried to include yews from around the UK and Ireland that are a varied mix, large historical ones, unusual individuals, woodland yews, hedges and topiary.

I have been asked many times which are my favourites. I couldn't possibly pick one, as they are all very special in lots of different ways.

Fossil records show that *Taxus baccata*, the species, is at least 15 million years old, long pre-dating ancient humans, who appeared around 2.5 million years ago, and modern humans (*Homo sapiens*). We have only been around for a mere 200,000 years.

Some yew trees are so old that they pre-date many of our Christian churches, whose churchyards tend to be strongholds for these ancient wonders. Venerated by pre-Christian religions, yews were often planted in groves to create places for meeting and worship, and are today surrounded in myths and mystery.

For me, my journey to visit some of these fascinating living icons has been an indulgence of my life-long fascination for trees, which extends from their natural history to their cultural history and the folklore that

Merrow Down yew

surrounds them. But no tree has as many myths and legends associated with it as our native yew. This is a tree that can outlive any other tree in Europe. Even the mighty English oak (*Quercus robur*), is outsurvived, by maybe as much as 2,000 or 3,000 years. The oldest living things in the UK are yew trees.

As a schoolboy, I loved to spend time in the woods during the holidays, getting a day rover pass for the Green Line buses to take me out of London. For an eleven-year old, heading off to places like Epping Forest for a day was quite an adventure!

More recently, visiting some of these ancient yews, with their vast hollow trunks, and learning how their cavernous interiors have been used as meeting places – some even fitted out with benches and doors – took me back to my childhood. During one school holiday, together with some friends, I discovered an absolutely huge old lime tree (*Tilia europaea*) that had a hollow as big as that of any tree I have ever seen (or that's my childhood memory of it at least!). The hollow wasn't at the base, but a short climb off the ground, and was like a ready-made treehouse. I remember there was ample room for five or six of us, and we would spend days in there, even camping in it overnight. We were joined by a family of starlings, the parent birds having chosen to raise their young in one of the smaller hollows. They seemed totally happy with this shared arrangement, going about their daily nesting and feeding chores as if we weren't there.

It was this early interest in trees that led to my first job after leaving school as an apprentice tree surgeon. Fittingly, old hollow trees became part of my life – this time elms. It was in the mid-seventies when many of the grand old English elms that furnished the countryside were succumbing to the devastating Dutch elm disease, and were being felled in their thousands.

The first of the great yews that I came across, more than 40 years ago, was the Ankerwycke yew in Berkshire, with its great fluted trunk and hollow centre. This tree is thought to be as many as 2,500 years old, but dating old yews is extremely difficult without historical evidence. Any suggested ages are at best guesstimates, based mainly on measurements of the trunk girth, usually at around 1 m (3 ft) from the ground, sometimes with historical information. But historical measurements can be unreliable, having been taken at different heights by different people, and many yews are difficult to measure because of their fragmented trunks.

Studying these trees has taken me to many remote parts of the UK and Ireland, to places that I would probably never have found or visited if it weren't for my quest to see these wonderful old treasures. Much of the landscape that surrounds them is wild and remote, and it has been a pleasure to spend time in such peaceful and stunning locations.

Previously I had thought that the yew was a tree of the southern chalklands, and that this was its preferred soil type and stronghold. But, as I now know, partly through writing this book, it is a tree that will grow in most soils. And in fact, a high proportion of ancient yews are to be found in Wales, rather than in southern England, though southern England does have the best remnants of ancient yew woods and forest.

Many of the magnificent ancient yews alive today can be seen in churchyards around the country, but these trees are thought to be pre-Christian, and so are older than the churches they stand beside. I am fascinated by these beautiful old churches and the churchyards in which they stand. Many contain original Celtic, Norman or Roman features, such as ancient standing stones, Roman tomb lids, and stones carved with early medieval Ogham lettering.

These old churchyards are also a refuge for wildflowers, looked after and maintained by local parishioners, volunteers and even secular charities such as Caring for God's Acre, which is dedicated to conserving and celebrating burial grounds. Some churchyards are home to over a hundred species of wildflowers, grasses and fungi, attracting a wide range of insects, including bees, butterflies and moths. These in turn attract other wildlife, such as slow worms and hedgehogs, both native species that are in decline.

Over the years, I have marvelled at the UK's wealth of ancient woodlands and trees, particularly oaks and yews, which are historically two of our most important trees. The oak's timber has been used for centuries in buildings and of course in the construction of Britain's early naval ships. But for me, it is our native yews that I find most fascinating, along with their significance to so many people and faiths, and the changes that the individual trees have lived through. Fortunately, many of these trees are well documented, particularly those in old churchyards.

Below The author and the Bettws Newydd yew

Right Layered branch

The yew has played a key role in religion, folklore, warfare and new cancer medication. This is very appropriate when you consider the yew's ability to regenerate itself, by producing clones where its lower branches touch the ground (layering) and by putting down aerial roots into its hollow centre, giving an ancient tree new life. This has made the yew a powerful symbol of life, death and resurrection.

The future of many of these amazing living relics is becoming increasingly uncertain, not only due to their great age and to climate change (as their shallow roots will not cope well with continued warmer, drier conditions), but also because they lack effective legal protection, as Tree Protection Orders are often flouted. Many have survived because they are on sacred ground, but even this safety may now be under threat. Dwindling congregations have led to many churches being sold off for development. As a result, some of these magnificent old trees, which have survived for so many centuries, could be at risk.

Some people believe that if their locations are kept secret, these trees will be safer. But I am convinced that the more people know about them and where they are, the safer they are. After all, it's much easier to get rid of a tree if no one knows it is there.

My hope is that by visiting these trees and spreading their stories, you will help to protect some of our amazing, natural, native living history.

NATURAL HISTORY OF THE YEW

Taxus baccata is known as the common yew, European yew or English yew. It is one of only three native conifers in the UK, easily distinguished by its very dark green, needle-like, forward-facing leaves set in flat ranks on spirally arching stems. The leaves are dark glossy green on their upper surface, and paler green with a raised midrib on the underside. The new stems stay green in their first two years, becoming brown usually in their third. The scaly bark is generally brownish, but has reddish and purplish patches and markings, and these colours really stand out and come to life when the trunk and stems are wet.

Common yew is an evergreen, small to medium-sized tree, generally growing to around 10–20 m (30–66 ft). But it can reach 25 m (82 ft) tall, or even more, in dense woodlands if it has to compete for light. It is dioecious, meaning there are separate male and female plants. Although rare, it has been known for parts of a tree to change sex. And it is usually just a single branch that unexpectedly does this. Most famously, the male Fortingall yew (p 104) in Scotland has a single female branch that produces berries. It is not fully understood why or how this happens.

Female branch on male yew

Yews can attain a great age, maybe becoming as old as 4,000 or 5,000 years, although more recent estimates are reducing these figures. These trees are, however, probably the oldest living things in Europe. Unfortunately, accurate dating is impossible. All ancient yews – those thought to be 1,000 years and older – have lost their heartwood through decay, leaving them hollow or filled with aerial roots and regenerated stems. To get an accurate date, you need a solid trunk, so that the annual rings can be counted.

Still, it is generally accepted that yews with a trunk girth of 9 m (30 ft) or more are at least 1,000 years old, and classed as ancient. But this dating method isn't totally reliable, as the girth of a trunk can vary significantly depending on the individual tree's growing conditions – especially the soil and aspect. In the south of England, on impoverished chalk soils, conditions are much harsher than, say, in fertile Welsh soils. And churchyard yews, which generally stand alone, tend to have a much wider trunk than trees growing in woodland, which are often taller with a smaller girth, as they have to reach upwards towards the light between other trees. Probably the harshest growing conditions of all are on steep cliffs, where yews can grow seemingly on almost pure rock, exposed to extreme weather, which restricts their growth. Along with many other experts, I believe that these cliff-dwelling yews may actually be our most ancient trees.

Wood anatomy

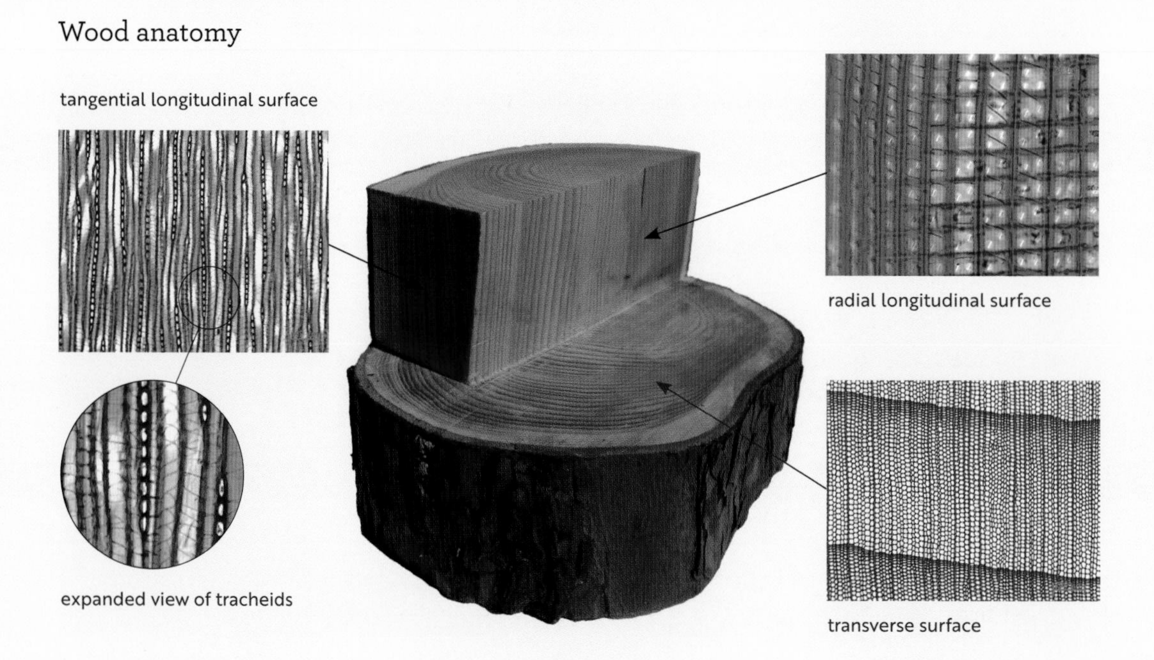

Pollen cloud

The yew is classed as a softwood, although unlike most softwoods, yew wood is actually relatively hard. Hardwoods are more complex in structure than softwoods – they contain vessels (hollow tubes to transport water), fibres for support, and axil and ray parenchyma for storage. Softwoods contain tracheids for transporting water and support, and do not have vessels. The three micrograph slide images on the page opposite show the anatomy of *Taxus baccata* wood, including the tracheids, or water-conducting cells, which are narrow and lignified with overlapping tapering ends. These transfer water from cell to cell via pits.

Along with other conifers and cycads, the yew is a gymnosperm, and as such it does not produce flowers. In their place are separate male and female reproductive structures. Yew (along with one of our two other native conifers, juniper) is referred to as a 'berry-bearing' conifer. Although yews do not produce a typical woody cone-like structure, unlike most other conifers and cycads in the division Gymnospermae, they do produce naked seeds (seeds that are not contained within an ovary). The term gymnosperm is derived from the Greek words *gymnos* meaning 'naked' and *sperma* meaning 'seed'.

On male yews, the small cones form in clusters on the undersides of the branches on short stalks. They produce vast clouds of pollen when fully ripe, from February to April. In woodland containing a lot of male yews, these clouds of pollen can be so huge that they have been mistaken for forest fires.

The yew's female cones, like the male cones, are produced on the previous season's shoots, and consist of a single ovule, which is the female reproductive structure that develops into the seed. This is surrounded by small overlapping bracts (modified leaves) at the base. The developing green female cones gradually turn into bright red (occasionally yellow), fleshy, cupped fruits, known as arils. These cover all but the tip of the poisonous seed hidden within, and are visible from late summer and autumn, usually persisting into January.

All parts of the yew, except the red fleshy aril, are poisonous to humans and most mammals, especially the leaves, which are lethally toxic, containing taxine alkaloids. Birds and some animals, such as badgers, do feed on them though. They can swallow the whole fruit, but only digest the sticky aril. The poisonous seeds pass through their system undamaged and are then spread via their droppings to more open areas away from the mother tree, where they can germinate more easily. Blackbirds, thrushes, fieldfares and redwings commonly feed on the arils, and in cold winters small flocks of these birds will descend on a tree to feast on them. Grey squirrels, too, enjoy the edible red arils.

Male cones

Young female cones

Red fruits

Yellow fruits

More recently, in southern England, I have noticed that the wild populations of ring-necked parakeets, which are originally native to India, have a real taste for yew fruits. But unlike most animals, which only digest the fleshy aril, discarding or excreting the poisonous seeds, the parakeets actually discard the aril in favour of the seed! I have observed them removing the red fleshy aril and cracking open the seed coating to eat the inside, and you can find the seed casings in large numbers beneath the trees. There is a yew species, the west Himalayan yew (*Taxus fuana*), that grows throughout the native range of this parakeet, so it might be that they have a natural defence to this otherwise poisonous seed.

Parakeets seem to have no problem eating the poisonous seeds

Taxus baccata has a much wider distribution than its common name of European yew suggests. Its geographical range includes Europe, western Asia and north Africa.

Other species of the genus *Taxus* are:

T. brevifolia – the Pacific yew, from western North America
T. canadensis – the Canadian yew, from North America
T. cuspidata – the Japanese yew, from north-east China, Japan, north-east Asia and Russia
T. floridana – the Florida yew, from Florida in North America
T. fuana – the West Himalayan yew, from Afghanistan, China, Tibet, India, Nepal and Pakistan
T. globosa – the Mexican yew, from Mexico
T. sumatrana – the tropical Asian yew, from Indonesia and the Philippines
T. wallichiana – the Himalayan yew, from northern Iran, Afghanistan, the Himalaya, north-east India and south-west China

Pests and diseases

Generally, yew trees are pretty much pest and disease free, as almost all of the plant is poisonous. The most toxic parts are the needle-like leaves, and curiously it is these that the caterpillars of the satin beauty moth (*Deileptenia ribeata*) feed on, along with the foliage of other conifers. These caterpillars are found from late August, and the adult moths are on the wing from June to August. They are not very common, and do not cause a serious amount of damage to yews.

A more widespread pest is the yew gall moth (*Taxomyia taxi*), and its presence is often very evident in the form of a gall that looks like a swollen terminal bud. This pest is exclusive to yews. It attacks the young shoot tips, causing the new growth to cluster into a rosette, which resembles a small globe artichoke or cone, rather than growing outwards as the shoot normally should. Within this cluster are the same number of leaves as there would have been if the shoot had continued to extend and grow normally.

The cone-like gall develops around the larva as it feeds inside the bud. Its entire life cycle until the point at which it emerges as an adult fly happens inside the bud, protected by the gall. Each gall contains a single, small, orange-red larva. The developing larva reaches pupa stage during April and May, and in June the tiny, 5 mm-long ($^1/_5$ in) fly hatches. The recently emerged female flies lay their eggs in the new young shoots, so starting the life cycle over again. The galls then turn brown, often persisting on the tree until the following year. Overall this pest causes no serious damage. I saw a particularly impressive display of artichoke galls on the Long Sutton yews (p 150).

Yew gall

Chicken-of-the-woods, or the sulphur polypore (*Laetiporus sulphurous*), is a bright, showy bracket fungus. It is a saprophytic fungus that feeds on dead and decaying hardwoods, and is frequently found on yews, usually growing at head height and above. It is often the first indication that a solid-looking tree is in the process of hollowing. Sulphur-yellow in colour (as its specific epithet *sulphureus* suggests), this fungus is commonly seen growing on the trunks of yews during summer and autumn. I found a good example of this bracket fungus growing on the Bettws Newydd yews (p 56).

Chicken-of-the-woods

Another group of bracket fungi, *Ganoderma* sp., are occasionally found on the decaying heartwood of yews. The bracket-shaped fruiting bodies tend to be seen low down on the trunk, usually below 1 m (3 ft).

Another disease affecting yew trees is caused by *Phytophthora* molds (from the Greek for 'plant destroyer'), of which there are around 60 known species. Many of these pathogens attack the tree's root system, destroying young roots, which reduces the plant's ability to take up water. Plants that are already under stress are generally more susceptible, which is often the case with many of our ancient yews. However, I saw very few yew trees affected by phytophthora during my research for this book.

DATING ANCIENT YEWS

Wherever there are ancient yews, there are claims that an individual tree is the oldest in the county, and in some cases the country. However, problems arise when trying to date these venerable trees accurately – the main reason being that as they move into old age, fungi begin to eat away at their heartwood, causing hollowing. This isn't always evident, but usually starts when a yew is around 400 years old. This decay makes the most accurate way of dating trees, by counting their annual rings, impossible, because the rings have disappeared.

But even if these trees were complete, with a solid, intact centre, the ring count on yews is still very complicated, as these trees have an irregular growth pattern. Their growth waxes and wanes throughout their life. The fluted trunks seen on many yews, particularly the older ones, are the result of this irregular growth in different parts of the tree. Often they also have long periods when there seems to be no growth at all.

Internal stem

Walking trees

There is a well-recorded tree at St Andrew's Church (p 192) in Totteridge, London, which because it has been growing from the inside, probably from an old aerial root or roots, has shown hardly any expansion in the girth of the main trunk for hundreds of years. In 1677, the girth was measured at 26 ft 3 in (8 m) by Sir John Cullum. It measured exactly the same in 1777, and more recently, in 1991, it was again measured at 26 ft 3 in (8 m) by Allen Meredith.

As these old giants lose their structural heartwood, they use other methods to ensure their survival and stability. They send down aerial roots to fill the internal void, and propagate themselves by layering – bending down their lower branches, which produce roots where they touch the soil, both stabilising the original tree and making young healthy clones. This process has given rise to the term 'walking trees', examples of these can be seen surrounding some of the woodland yews in Kingley Vale (p 130). The energy of the tree is diverted into this new growth, at the expense of expanding the girth of the main trunk.

Carbon dating the oldest trees is not a feasible option either, as this also relies on analysing the oldest wood, which has decayed and disappeared.

Historical records of these trees tend to be patchy. For quite a few old yews, particularly those in churchyards, there are usually some records, but these frequently only go back to the 17th century and often state that the tree was already ancient at that time. It would also seem that although many of the villages and some of the churches are mentioned in the Domesday Book of 1086, no individual trees are recorded there, from churchyards or elsewhere.

The method currently used to date most ancient yews is still at best just a guesstimate. It is based on the girth measurement of the tree at a given height, most commonly at 1 m (3 ft) or as close to that as possible. Many trees have also been measured at ground level, and some at both points. But as this measuring has been done by different people over the years, even this may not be consistently accurate. Some individual trees have been measured repeatedly over more than 300 years, including the Crowhurst yew (p 72) in Surrey. Its first recorded measurement was in 1630, with a 30 ft (9.15 m) girth, and there are records through the intervening centuries up to modern times.

Fragmented trunk

Trees that have fragmented trunks are often very difficult to measure accurately, because the fragmented sections tend to lean outwards, away from each other, probably giving a larger girth measurement than if the trunk had stayed in one piece. A tree may also produce burrs, bulges and epicormic growths around the measuring area, which can lead to a misleading increase in the girth.

Another problem I see with this method of guesstimating the age of a tree (and I am not sure there is currently a better one) is that the growth rate of all trees depends partly on the environment they are growing in. For instance, yews that grow on the dry chalklands of southern England will have smaller incremental growth than, say, that of yews growing in the wetter parts of Wales. While for upland and cliff-dwelling yews, the growing conditions can be extremely harsh and growth is so slow that it takes a microscope to distinguish between the different growth rings.

Some accurate ring counts have been carried out on the solid parts of certain yew trees, for instance a fallen branch from the crown of one of the Borrowdale yews (p 60) has been dated by scientists to 1,500 years old. The whole tree is therefore older than this.

So a reliable, accurate measurement record, plus historical information, backed up with good photographic images, is the best information that we have about the ages of yew trees at the present time.

RELIGION AND FOLKLORE

Religion

Most of the oldest yew trees in the UK are found in churchyards. There are lots of theories as to why they are there: perhaps because the walled enclosure of a churchyard stopped livestock from eating their poisonous leaves and bark, or conversely because these poisonous trees discouraged farmers from bringing their livestock into the churchyard to graze, or because the yew was protected in the churchyard, so its timber would be available for longbow-making. I think the last suggestion is unlikely though, due to the quantity of suitable timber needed to make bows – woodland and forests would have been more likely sources. It is also well documented that much of the yew wood used to make English longbows came from overseas, particularly from France and Spain.

For me, the most convincing reason is that they were deliberately planted on sacred and religious sites. We know this occurred as far back as the Bronze Age – the mound on which the ancient yew in the churchyard of St John the Baptist (p 48) in Ashbrittle, Somerset, was planted is believed to be a Bronze Age barrow.

The pre-Christian priests or druids, although they may or may not have worshipped the trees themselves, are thought to have used yews as places to gather. They also planted trees to form groves, or used natural groves in which to worship. And there are instances where there are circles of yews that surrounded churches, particularly in Wales. I wonder whether many of the old circular churchyard boundary walls were built that way to enclose a circle of ancient trees, now long gone, and whether there is also a link between yew tree circles and ancient stone circles as places of worship. Famous stone circles like those of Stonehenge and Avebury in Wiltshire, whose original purpose is unknown, are thought to have been religious sites and most likely used for rituals and ceremonies.

It is believed that ancient pagans, who were nature worshippers, did venerate trees, and some modern pagans do so today, as shown by the offerings you find left in these ancient trees, particularly around the summer and winter solstices. It is also well documented that yew trees were associated with longevity and fertility. It was believed that the yew was particularly sacred, being an evergreen tree, in leaf the whole year round and long associated with rebirth. This may have been because of the

Offerings left by modern pagans

yew's ability to re-grow from within its hollow centre by producing aerial roots, to propagate itself by growing new clones where its lower branches touch the ground, and to produce new growth from old wood.

Many of the churchyards where ancient yews grow were originally sites of Druidic or Celtic worship, and later were the sites of early temples of worship and later Anglo-Saxon churches. They continued to be sacred places when the pagan population was converted to Christianity, from Roman times onward. Newer churches, built over the old ones, were re-dedicated to Christian saints. Several yews remain from those pre-Christian times.

Yews continued to be of great importance to the Christian churches. In the UK and northern Europe – where palms didn't naturally grow – the tree often played a central role in Palm Sunday celebrations, commemorating Jesus's entry into Jerusalem just before his crucifixion. During medieval times, Palm Sunday was also also known as Yew Sunday.

The link between churches and yews became so established that when new churches were built, yew trees were often planted next to them.

Mythology, legends and folklore

Ancient yews, both churchyard trees and those in woodlands, do have an air of mystery about them, even in this day and age, when we have a better understanding of science than our ancient predecessors. So it is easy to imagine the wonder and perhaps even fear that these trees could conjure up thousands of years ago. A lot of these trees are now 'managed' and would have looked very different in the distant past, with more branches twisting their way down to the ground around the main trunk, often creating an almost impenetrable, fortress-like wall of branches.

I remember when I first saw the old yew in Holeslack Woods (p 120) in Cumbria, how eerie it felt coming across this beautiful, impressive tree, deep in the woods on a dark day. While the trees at Kingley Vale (p 130) in Sussex are otherworldly, with twisting branches full of dead wood forming tunnels that lead you into almost cathedral-like spaces around their trunks. Some are short and squat, others much taller, reaching up for the light, or leaning at angles on their grounded branches as if taking a rest. How different these many-branched trees are to the often pruned churchyard yews. It is no wonder that they inspired mystery and legend.

Yggdrasil, the mythological tree of life, was originally thought to be a giant ash tree. Ancient sources, including the Eddas (a collection of Old Norse poems), mention *vetgrønster vida*, meaning evergreen tree, and an Old Norse word, *barraskr*, meaning needle ash. These both suggest that the *Yggdrasil* is evergreen and needle bearing. An ash is neither of these. So it is thought *Yggdrasil* was most likely a giant yew.

Druids believed that wands of yew would banish evil spirits, bringing purity and peace. And yew staves carved with early medieval Ogham text were used to perform magical tasks. The waters from wells beneath or close to yew trees, such as that near the Hope Bagot yew (p 122) in Shropshire, were also thought to have magical healing properties.

The famous Bleeding Yew (p 58) in Nevern, Dyfed, oozes a red substance from a wound on a cut limb, looking very blood-like, and is linked to many myths. For example, the tree is said to be bleeding to protest the innocence of a young monk, wrongly hung in this tree. The monk supposedly vowed with his last breath that the tree would bleed for all eternity, as a reminder of his innocence.

The yew was a symbol of death and resurrection in Celtic culture, due to its ability to re-sprout and put on new growth after years of inactivity. This tree is also said to symbolise death and resurrection in Christianity.

Yews marked the end of the year in the Celtic calendar, and the autumn festival of Samhain is one of the most important dates in the pagan calendar. We now celebrate it as Halloween. It was believed to be a time when spirts could more easily cross from the Otherworld into ours, falling between the autumn equinox and the winter solstice.

Long associated with magical abilities, death, rebirth and eternal life, the yew has played its part in literature. Several of Shakespeare's plays make references to the yew. In Macbeth, a poisonous potion being concocted by the three witches includes the ingredients 'scale of dragon and tooth of wolf' as well as 'slips of yew silvered in the moon's eclipse'. In Twelfth Night, the fool, Feste, sings of his 'shroud of white, stuck all with yew'.

Many poets, including William Wordsworth and Alfred, Lord Tennyson, have written about the yew. Tennyson symbolises the yew tree as having the power of death in his poem *In Memoriam – 2.*

Old Yew, which graspest at the stones
That name the under-lying dead,
Thy fibres net the dreamless head,
Thy roots are wrapt about the bones.

The seasons bring the flower again,
And bring the firstling to the flock;
And in the dusk of thee, the clock
Beats out the little lives of men.

O, not for thee the glow, the bloom,
Who changest not in any gale,
Nor branding summer suns avail
To touch thy thousand years of gloom:

And gazing on thee, sullen tree,
Sick for thy stubborn hardihood,
I seem to fail from out my blood
And grow incorporate into thee.

Tennyson's yew

USES

Yew wood is among the hardest of all softwoods, yet it is extremely flexible and very strong, with a fine, tight, straight grain. These qualities make it an ideal material for many applications. The world's oldest-surviving wooden artefact is a yew spearhead found in Clacton-on-Sea, Essex, in 1911, which is estimated to be around 400,000 years old.

Archery

What of the bow?
The bow was made in England;
Of true wood, of yew wood,
The wood of English bows;
So men who are free
Love the old yew tree,
And the land where the yew tree grows.

Sir Arthur Conan Doyle

Historically, yew was the preferred wood for making longbows. Other woods were used, particularly elm, but a bow made from yew was far superior. Both the heartwood and the sapwood were used. The inner side or belly of the bow is the orangey-brown heartwood, which resists compression, while the outer creamy-white sapwood works in tension. This natural laminate gives this style of bow tremendous flexibility and power, making it able to fire arrows a distance of well over 200 m (650 ft).

Bows were used for both hunting and warfare. The longbow is a well-documented weapon of the Normans and was used with devastating affect in the Battle of Hastings. Throughout the medieval period, this powerful weapon decided many battles, including several during the Hundred Years' War of the 14th and 15th centuries.

During this period, one of the major conflicts in which the English longbow was a decisive weapon was the Battle of Agincourt. The 7,000 longbow men – both English and Welsh – formed a large part of Henry V's army. Although the English forces were greatly outnumbered, the French were defeated, in the main by the volleys of arrows from the English army, which inflicted fatal injuries on the French cavalry and men-at-arms.

In the yew bow's heyday, it was decreed that only the royal longbow men could have yew bows, partly because good-quality yew wood was in high demand and becoming scarce. So much yew forest was depleted of suitable timber that yew wood for bows had to be imported into England as early as the 13th century.

Excitingly, 137 complete yew longbows and more than 3,500 arrows were found on the Tudor warship *Mary Rose*, which was discovered on the seabed in the Solent in 1971 and raised in 1982. This find has allowed much experimental work to be carried out on the centuries-old weapon. Some of the longbows are on show in the Mary Rose Museum in Portsmouth.

The oldest yew bow ever found in the UK was discovered during commercial peat extraction in the Somerset Levels in 1961 and is known as the Meare Heath bow. The bow – or more accurately half of the bow – with its distinctive long, flattened shape, still has a section of the hand grip remaining, as well as part of the terminal notch used for attaching the bowline. Two leather bands were fixed around the bow-stave, and there are traces that show where other banding strips would have been attached along its length, as well as the remains of diagonal markings, thought to have come from sinew used to add strength to the bow. The bow has been dated to the Middle Neolithic Period, around 2,690 BC.

Wood turning and furniture making

Traditionally, yew has been used in wood turning, mainly to make bowls and cups or mugs, but also furniture legs and spindles.

Being distinct from other woods, and particularly beautiful due to its rich colouring, yew has been used for making furniture for hundreds of years. The creamy-coloured sapwood contrasts with the orangey-brown heartwood, and is much prized by cabinet makers for making exceptionally decorative veneers. The tiny 'pippy' knots and burr wood, with their unique patterning, are highly sought-after to make veneers in high-quality furniture and cabinet making. It was traditionally believed that beds made from yew wood would deter bed bugs.

Musical instruments

The world's oldest surviving wooden instrument is made up of six yew-wood pipes, which were discovered during an archaeological excavation in 2003 in Co. Wicklow, Ireland. They were found by archaeologist Bernice Molly in a waterlogged trough from the Early Bronze Age (c. 2,200–2,000 BC). The sections were laid out side by side in descending order of size. They were hollowed out, but had no finger holes, so it is thought they may be part of an unfinished wind instrument.

A beautiful yew bowl turned by Jake Davies-Robertson

Yew was one of the favoured European woods for the pear-shaped back body of lutes, which were constructed from an odd number of thin strips of wood (commonly nine or eleven in early 16th century lutes). These were bent and fixed over a mould, with their edges glued together, to form the shell-like structure. Many of the multi-rib lutes that survive from this period are made from yew.

Topiary

Yew has always been a popular plant for hedging, particularly for its ability to withstand being pruned or cut back continually. This made it ideal for topiary (the art of clipping evergreen plants into formal, stylised shapes), which was very popular during the Italian Renaissance (14th–16th centuries) and in the 17th century in England. The gardens of Levens Hall (p 136) in Cumbria still feature original topiary that dates back 300 years.

But the art of topiary goes back much further and has been practised in Europe since Roman times. Pliny the Younger, for example, describes in a letter the various forms and shapes that shrubs were cut into in his gardens at Laurentum, close to the sea just west of Rome.

Right Harlington topiary yew today

Below Harlington churchyard old topiary yew

Even some English churchyard yews have been clipped as topiary. There is a large yew tree in a churchyard in Harlington, in the London borough of Hillingdon, that was clipped into an intricate shape in the early 18th century, said to be 15–18 m (50–60 ft) tall. This tree had ceased to be clipped by the end of that century and gradually reverted to its natural shape.

Topiary remains popular today, and although yew is still one of the favoured shrubs, there are many other plants that are now widely available and will also take regular clipping into intricate shapes, among them common box (*Buxus sempervirens*) and box honeysuckle (*Lonicera nitida*).

Chemicals

The Pacific yew (*Taxus brevifolia*), a species from North America, has been used by indigenous people to make medicinal leaf preparations to induce healthful sweating, and it is said that leaf tea was used to treat lung ailments. This species was also known as the bow plant – like the European yew, it was used to make bows and arrows. Canoe paddles, harpoon shafts and even fishing hooks were also made from it.

It was also from this species that scientists in the 1960s first discovered yew's potential to treat cancer, using a natural compound found in its bark, called paclitaxel (sold under the brand name Taxol). This is an antimitotic agent used to block cancer cell growth.

Another anti-cancer drug, docetaxel (Taxotere) was first made from the needles of the European yew tree.

The increasing demand for the compound in subsequent decades had a devastating effect on the population of the species. Fortunately, a synthetic form of the Taxol compound was developed in the early 1990s. Taxol still remains one of the most effective plant-based treatments for cancer.

Roman author Pliny the Elder noted that people had died after drinking wine stored in barrels made from yew. Yet wooden staves cut from yew were used in Ireland for making wine barrels, it seems with no ill effects!

CHURCHES AND CHURCHYARD FEATURES

Many of the churches I have visited still show signs of their ancient past within their structures, such as Saxon foundations, Norman towers and decorative arched doorways. Even after renovations and modifications (and, in some cases, complete re-builds) over the centuries, often carried out during the Victorian period, some still show fascinating links to their early beginnings.

One of the joys of visiting the yews that grow in association with churches are the churches themselves, and the history that surrounds them. Travelling around the country to visit yews, you will find the churches extremely variable in their locations, ages and styles, from small simple chapels to huge Victorian edifices.

Below St Mary's single-celled church, Pontrhydfendigaid, Ceredigion

Opposite St Winifred's standing stones, Gwytherin, Wales

Above St Michael's in Llanfihangel Nant Melan, Powys

Below Red sandstone, St Andrew's, Kenn, Devon

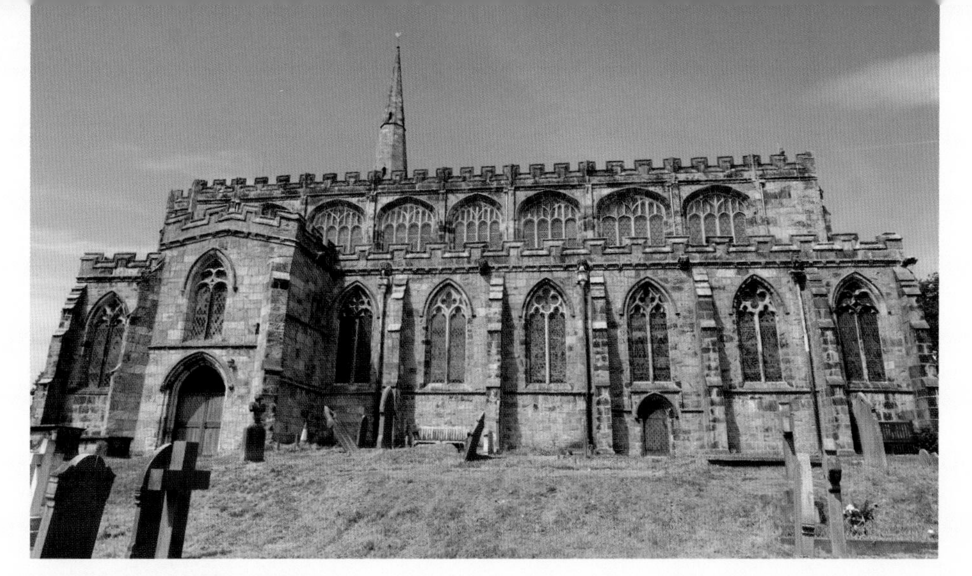

Above St Mary's, Astbury, Cheshire – a large Victorian church

Below North Downs flint, St Edmund's, West Kingsdown, Kent

Some churches, such as St Michael's in Llanfihangel Nant Melan (p 140), appear to be older than they are. This church is Norman (11th/12th century) in style, but was actually built by the Victorians in 1846.

Churches are usually built from local, easily accessible materials, particularly the early churches, like the beautiful red sandstone and white Beer stone from which St Andrew's Church (p 128) in Devon is built, and the flint and chalk-rubble walled churches in Kent, such as St Edmund's in West Kingsdown with its local stone from the North Downs.

Inside many churches, Saxon and Norman fonts may be seen. Often these are all that remains of the earliest church on the site. Sometimes you will see original timbers in roofs and rood screens dating back to these periods, as well as amazing wall paintings, like the one below.

Wall painting of Medieval jousting knights, All Saints' Church, Claverley, Shropshire

This wonderful frieze, at All Saints' Church in Claverley, Shropshire, dates from 1220 and is considered to be one of the best examples of medieval wall art in Britain. It was discovered under whitewash during renovations in 1902. The jousting knights in the frieze are thought to represent battles between Virtues (the gold horses) and Vices (the red horses), although there are other interpretations. The paintings show riders in chain mail on different brightly coloured horses, fighting with shields, swords and lances, with many of the riders being unseated during this stylised painted battle.

Many of the churches reveal features that are older than the main structure of the building. I have seen an amazing array of fonts, some from an earlier religious structure than their current ones. They are often the oldest remaining remnant, and the church's origin can be dated by these features. Saxon fonts tend to be fairly plain, but the elaborate limestone Norman font pictured is decorated with carved columns of different designs, joined by arches with stylised leaf patterns above.

Left 12th century Norman font

Below Saxon font

Dating from the 11th century, Wilmington Priory is now an atmospheric ruin. Nearby stands the Wilmington yew (p 206), in the churchyard of the local parish church.

Similarly, close to the yew at Ankerwycke (p 46) are the ruins of what was once a nunnery, dedicated to St Mary Magdalene and built during the reign of Henry II. After the Dissolution of the Monasteries under Henry VIII, it passed into private ownership. Today, all that is left are small ruined remnants of the old nunnery's walls. Abbey and monastery ruins are fairly common neighbours to old yews and churches.

Left Wilmington Priory, Polegate, East Sussex

Opposite left
10th-century grave marker

Opposite right
Stone coffin lids

Many early sacred sites were later adopted as Christian places of worship, along with any yew trees growing there. You will often see features in churchyards that suggest they have been spiritual locations since pre-Christian times, such as standing stones, Roman gravestones, Celtic crosses and stones with early medieval Ogham text inscriptions.

Ancient gravestones and interesting grave markers can be seen in many churchyards, sometimes in strange locations. One curious example can be seen at St Mary and St Peter's Church (p 206) in Wilmington, East Sussex – here, a Roman stone discovered at the bottom of a vicarage well has been re-used as a grave marker for the deceased well-digger who originally found it. You can see this stone at the base of an ancient propped yew, which has started to envelop it.

Some old gravestones and markers occasionally turn up in the church walls themselves, having been recycled as building materials long ago. Two such stones were found in the wall at St Brynach's Church (p 58) in Nevern, Wales, one with an unusual knotting pattern in the shape of a cross, and the other with incised early medieval Ogham text. Both are now displayed on the internal windowsills of the church.

There are four standing stones in the churchyard of St Winnifred's Church (pp 28, 108) in Gwytherin, Wales, of an unknown date. They run in a line east to west along the northern side of the churchyard's boundary, and are more or less evenly spaced. The westernmost stone has an incised Latin inscription on it that reads: VINNEMAGLI FILI SENEMAGLI. This is thought to translate as 'Vinnemaglus, son of Senemaglus', and dates to the fifth or sixth century AD. It may have been added to this stone at a later date.

Many churchyards are wonderful havens for wildlife, being carefully managed to encourage native flora and fauna. Many of the flowers that grow in them and the animals that use them as a refuge are becoming endangered. These include the hedgehogs, which were once numerous and are now a rare sight, and slow worms, which used to be common in any garden with a compost heap or old boards lying around to provide a hiding place. Again their numbers are greatly reduced through lack of habitat. Churchyards are often home to an abundance of native wildflowers throughout the year, attracting insects, which in turn attract birds. Butterflies and bees come for the nectar-rich wildflowers. A late spring visit to these 'church meadows' is truly a naturalist's delight.

Hope Bagot, Ludlow, Shropshire, is managed for wild flowers

The drystone walls and hedgerows that mark churchyard boundaries are another asset, providing nesting and shelter for many living things. And in autumn, berries and other hedgerow fruits help animals to stock up their reserves for the winter ahead.

When I was a child, many of these wild creatures were a familiar sight, including sparrows, starlings, linnets and greenfinches. Feeding hedgehogs in the garden and catching slow worms after school are no longer commonplace for children. Among the hustle and bustle of modern life, churchyards are wonderful places in which to stand and draw a breath, and take in our wonderful natural history.

My primary intention was to visit some of the great ancient churchyard yews, but I have seen far more trees and amazing flora and fauna than can be included in this book. I hope that during any visits you make to these churches, you will take the time to have a good look around and enjoy the other treasures that surround them.

CHURCHYARD YEWS

Our oldest churches, and churchyards, are a stronghold for ancient yews. There are many opinions as to why ancient yews in particular are so closely associated with holy places. After all, English oaks can attain a grand old age as well – a thousand years, although rare, is not unheard of. However, many yews far surpass that age, and it is almost always old yews, and not oaks, that are associated with ancient religious sites.

Some pre-Christian faiths worshipped the natural world and the elements, including the sun, moon and trees. This may be one reason why some of the ancient yew trees were planted in special places or sacred sites, which have since become today's churchyards.

The yew, in particular, was worshiped by the pagans for its links with death and re-birth, as an emblem of resurrection. This tree was also sacred to the Druids (the priests of the Celtic tribes), who erected temples close to them. As evergreens, yew trees were a symbol of everlasting life, and the twigs and leaves were laid on graves to remind the departed spirit that death was only a rest in life, before rebirth.

St Aeddan's Church, Bettws Newydd

St Mary Magdalene, West Tisted, Hampshire

Christianity began to arrive in Britain along with the Romans during the third century AD and many of the pagan sacred sites were adopted by the new religion. This is why many of the oldest yews pre-date the churches they stand alongside. By the Anglo-Saxon period (400 AD to 1066), Christianity had began to dominate. During Tudor times, it was customary to tie sprigs of yew to coffins, and yew has long been used as a church decoration during Palm Sunday celebrations, so it would have been convenient to have a regular supply on hand.

In addition to their sacred significance, yews are thought to have been planted within churchyards so that livestock wouldn't be able to eat the toxic leaves. Alternatively, they were perhaps planted there to prevent farmers from grazing their livestock on holy sites.

Yews have also long been planted as 'consecration yews' alongside Christian churches, and this is a custom still widely practiced in today, particularly during church and churchyard renovations. Often you will find one yew at the lychgate (the funeral entrance to the churchyard) and another close to the church door. The upright Irish yew is also often planted along the pathway leading to the church door. Some churchyards contain many yews, both common and Irish. St Mary's Church (p 162) in Painswick, Gloucestershire, has more than one hundred, but this is exceptional.

THE YEWS

Location of yews featured in this book

No.	Name
1	Abbotstone Down Yews
2	Aberglasney Yew Tunnel
3	Ankerwycke Yew
4	Ashbrittle Yew
5	Astbury Yew
6	Bathurst Estate Yew Hedge
7	Benington Yew
8	Bettws Newydd Yews
9	Bleeding Yew (Nevern)
10	Borrowdale Yews
11	Breamore Yew
12	Claverley Yew
13	Coldwaltham Yew
14	Crom Castle Yews
15	Crowhurst Yew (Surrey)
16	Crowhurst Yew (Sussex)
17	Cudham Yews
18	Cyffylliog Yews
19	Darley Dale Yew
20	Defynnog Yews
21	Didcot Yew
22	Discoed Yews
23	Doveridge Yew
24	Downe Yew
25	Druids Grove Yews
26	Dunsfold Yew
27	Eastling Yew
28	Farringdon Yews
29	Florence Court Yew
30	Fortingall Yew
31	Goostrey Yew
32	Gwytherin Yews
33	Hafod Yews
34	Hambledon Yews
35	Hampton Court Palace Maze
36	Holeslack Wood Yew
37	Hope Bagot Yew
38	Iffley Yew

No.	Name
39	Itchen Abbas Yew
40	Kenn Yew
41	Kingley Vale Yews
42	Langley Yew
43	Leeds Yew (Kent)
44	Levens Hall Topiary Yews
45	Linton Yew
46	Llanfihangel Nant Melan Yews
47	Llangattock-juxta-Usk Yew
48	Llangernyw Yew
49	Llansilin Yews
50	Lockerly Yew
51	Long Sutton Yews
52	Lorton Yew (Wordsworth's Yew)
53	Martindale Yew
54	Much Marcle Yew
55	Newlands Corner Yews
56	Painswick Yews
57	Payhembury Yew
58	Pennant Melangell Yews
59	Pulpit Yew
60	Rock Walk Yews (Wakehurst)
61	Seatoller Yew
62	Silken Thomas Tree
63	Staunton Yew
64	Stedham Yew
65	Stoke Gabriel Yew (Devon)
66	Stow-on-the-Wold Yews
67	Strata Florida Yew
68	Tandridge Yew
69	Tisbury Yew
70	Totteridge Yew
71	Ulcombe Yews
72	Watton-on-Stone Yew
73	Waverley Abbey Yew
74	West Kingsdown Yew
75	West Tisted Yew
76	Wilmington Yew

30
52
61
10
53
29
14
36
44
62
48
59
18
31
5
19
23
32
58
49
12
33
67
37
22
9
46
54
45
2
20
63
56
66
6
38
42
21
3
47
8
72
7
70
17
24
74
43
35
51
55
27
71
28
1
15
4
57
68
40
39
41
64
25
60
16
13
76
69
11
50
34
26
75
73
65

Abbotstone Down Yews

1

Location. Abbotstone Down Nature Reserve, Near Old Alresford, Hampshire SO24 9UD; off the B3046 (between Northington and Old Alresford).
OS grid ref. SU582362 (For directions to the trees, see below)

Wild things

Abbotstone Down lies just to the north of the village of Old Alresford and was formerly an area of chalk grassland, of which only remnants now remain. Much of it is now covered with young woodland, although there are a few mature oaks. The area also includes the remains of a small Iron Age hill fort.

Within the wooded areas, which are mainly made up of young ash, are some scattered yew trees, none of which would be classed as ancient, but they are old nevertheless. They are spread across this hilltop, with some in small clusters.

The trees are mostly multi-stemmed, with a fairly low (around 10 m/ 30 ft high), spreading habit. The limited height is due to the fact that they sit on the high ground of the downland and are sparse trees that have been growing in mostly open grassland, with no need to be tall (unlike those in woodland that have to reach for the light). Instead, they have branched from low down, spreading wide, with their lowest limbs rooting into the ground. These unpruned wild forms can be almost impenetrable, with a fortress of branches hiding their trunk.

The varied shapes of these yews are now, however, shrouded by small trees and scrub. The low light levels also create an almost eerie atmosphere, adding an extra air of mystery to these already strange trees.

Directions To find the main group of trees, follow the footpath from the small carpark, past the fenced-off area on your right, until just after the third set of wooden kissing gates, and turning right into the woodland. A few minutes' walk in through the hazel, ash and hawthorns are some of the larger yews, in a grove of a dozen or so trees made up of both males and females.

In the fenced-off compound closer to the carpark are some other yews within a post-and-railed area, erected to keep grazing cattle away from the poisonous yew foliage.

Originally in open grassland, the wild yews of Abbotstone Down have grown low and wide, as they weren't restricted by other trees nearby

Aberglasney Yew Tunnel

2

Location. Aberglasney Gardens, Llangathen, Carmarthenshire SA32 8QH.
OS grid ref. SN579221

Tunnel vision

The impressive 30 m (100 ft) long yew tunnel in Aberglasney Gardens is thought to be a unique feature in the UK. It was obviously a planned feature, and was probably originally a yew hedge, with the branches being deliberately bent over at some later stage to form the tunnel we see today. Over time, the individual yews have become so entwined with each other that it is now difficult to establish how many trees make up the tunnel, but it is believed to be around half a dozen.

The tunnel was once thought to be extremely old, maybe as much as 1,000 years, and to be part of the original medieval gardens that surrounded the 16th-century house. But work by dendrologists in 1999 to establish the age of the yews has shown them to be around 250 years old, so unfortunately not as ancient as once assumed. This means that they may have been planted by the Dyer family in the 18th century, or a little later by the Phillips family.

The Victorians loved to visit 'pleasure gardens' like Aberglasney to see the many new plants that were coming into the UK at that time. And curiosities like the yew tunnel would have been enjoyed, well visited, carefully maintained and regularly clipped. But it became neglected in the mid-20th century, when maintenance stopped and the branches grew long, upwards and outwards. In recent years, however, all of the neglected growth has been cut back, returning the tunnel to its full glory.

The 1999 BBC television series *Aberglasney: a garden lost in time*, which followed the restoration of the garden, including the yew tunnel, brought the estate to public attention. It is now one of Wales's finest gardens.

Probably originally planted as a hedge, the yews were deliberately bent over at some stage to form what is now a unique 30 m (100 ft) long tunnel

Ankerwycke Yew

3

Location. On National Trust land near Wraysbury, Middlesex TW19 5AD.
OS grid ref. TQ004726

Magna Carta connections

I first came across this magnificent yew thanks to my love of fishing. This is the tree that sparked my interest in old trees, and particularly in ancient yews and the myths and history surrounding them.

In my early teens, I used to go fishing at weekends with my friends, making a lengthy bus journey from my home in Ealing to the Thames at Runnymede. One unproductive day, as I walked along the riverbank to see how the others were doing, I came across signs for the Magna Carta Memorial. Intrigued, I crossed the road and walked up the hill to see this monument, where I found out about the historic events that took place here, and the nearby ancient yew tree.

In Saxon times, Runnymede was known as Rune-mede, or meadow of the runes, where councils would meet to consult the runes. And it was in this area in June 1215 that the Magna Carta's oaths and the sealing of the act were performed by King John and the barons of England. The exact site where the signing took place is not known for sure, but some historians believe it was beneath the ancient tree known as the Ankerwycke Yew. Richard de Montfichet, who owned the manor of Ankerwycke in which the tree stands, was one of the barons present at the ceremony.

The Ankerwycke Yew is thought to be up to 2,500 years old, making it the oldest known tree on National Trust land, and it stands as a lone male yew on a high point in the flood plain, across the river from Runnymede. Its great fluted trunk, like those of so many other ancient yews, is hollow in the centre, but has begun to refill with lots of aerial roots, probably at the cost of extra girth growth but hopefully strengthening the tree's future. These aerial roots grow down into the ground and become thicker over time, eventually turning into stems in their own right. One of the tree's branches has also rooted and is producing new growth – making it what is sometimes known as a walking tree.

Not far from this tree are the remains of St Mary's Priory, a Benedictine convent built in the 12th century in the reign of Henry II. The yew would have already been a notable ancient landmark, substantially more than 1,000 years old, when the priory was built. Legend also has it that Henry VIII courted the ill-fated Anne Boleyn beneath this tree.

In more recent times, the Ankerwycke Yew was granted a Tree Preservation Order in April 1990, and in 2002 it was chosen as one of the '50 Great British Trees' to celebrate the Queen's Golden Jubilee.

The Magna Carta may well have been signed beneath the Ankerwycke Yew, 800 years ago – at that time it was already an ancient tree, some 1,700 years old

Directions In the small carpark at the end of Magna Carta Lane there is a map with directions to the tree and priory ruins.

Ashbrittle Yew

4

Location. In the churchyard of St John the Baptist Church, Rectory Lane, Ashbrittle, Somerset TA21 0LQ. **OS grid ref.** ST051213

One of the all-time greats

This venerable yew in the small Somerset village of Ashbrittle is thought to be among England's most ancient survivors. The plaque alongside it declares: 'Generations of local people have cherished this tree, one of the oldest living things in Britain'. It also proudly claims that the tree was already mature when Stonehenge was in use and is reputed to be more than 3,000 years old. It is certainly the largest yew in the county of Somerset.

As with many other ancient and veteran yews, it stands among the gravestones of an old churchyard. But step back and you will notice that it also grows on a mound. This raised area, just to the south east of the church, is believed to be a Bronze Age barrow. According to local legend, the church is built on the site of a druidic circle, which is quite plausible as sacred sites are often used repeatedly throughout history as places of worship. There is also a spring close by, which now flows under the altar of the church – traditionally, yews were planted next to springs as symbols of everlasting life.

This tree has a huge canopy of twisted branches, some of which are in decline. The gnarled trunk is now in seven pieces, and the girth of this fragmented trunk has been measured at more than 12 m (40 ft). There is a large, hollow central trunk, with six smaller ones forming what may have been the outer edge of the original tree. The central trunk is probably the result of the tree trying to repair itself long ago by putting down aerial roots in the centre, as this part decayed and rotted away.

The Ashbrittle Yew was chosen as one of the '50 Great British Trees', to mark the Queen's Golden Jubilee in 2002.

Standing on a mound, probably a Bronze Age barrow, the yew has seven trunks and a vast canopy that overshadows several gravestones

Astbury Yew

5

Location. **In the churchyard of St Mary's Church, Astbury, Congleton, Cheshire CW12 4RQ. OS grid ref. SJ845615**

Teetering on the edge

To call this yew hollow would almost be an understatement. It is literally just an old shell of a tree, leaning over the path and propped up on a wooden telegraph pole and a large purpose-cut forked branch. But the ability of old yews to continue growing from almost nothing never ceases to amaze me.

This old male yew is probably just part of a long-ago fragmented tree, now just surviving in the two sections that are left. The living parts of this tree are only around the extremities. Some of these are starting to callus over the old dead wood, giving the trunk a mystical, timeless feel, with the old silvering dead wood contrasting with the living dark-red wood.

Looking up into the crown, you will be impressed by its health, considering what is left to sustain it. It seems able to produce new growth all over its upper shoots, creating a healthy-looking crown.

The churchyard itself contains many other yews of varying ages and sizes, including fastigiate (upright) and clipped yews, but no others that have been around for as long as this propped-up old gentleman.

With such a tattered and toppling trunk, mainly made of dead wood, it's amazing the tree has such as healthy-looking canopy

6

Bathurst Estate Yew Hedge

Location. Bathurst Estate, Cirencester, Gloucestershire GL7 2BU.
OS grid ref. SP020020

High-rise hedge

A spectacular sight, this is the tallest yew hedge in the UK, and probably the world. It is thought to date back to the early 18th century, which would mean it was planted around the time that the existing mansion was built.

The 300-plus-year-old hedge is approximately 12 m (40 ft) tall, 4.5 m (15 ft) thick and nearly 140 m (450 ft) long. Unusually, it is semi-circular in shape, following the curve along the front boundary wall of the mansion, screening it from the town of Cirencester. If you climb the tower of St John Baptist Church in the Market Place, you get an unobstructed view of this tremendous hedge and the estate beyond it. I wonder if the early intentions of the first Earl of Bathurst or the gardeners in the 18th century were for a hedge on quite this vast scale?

The hedge is trimmed annually, usually in August, and this would originally have been done standing on tall and probably rickety ladders, making it a rather precarious ordeal! Today it is done more safely from a large cherry picker. The trimming of this monstrous yew hedge takes over two weeks and produces around one tonne of clippings. These are collected and donated to the pharmaceutical industry to produce an extract from the yew leaves that is an ingredient in the chemotherapy drug docetaxel, which is used to treat various forms of cancer.

Unfortunately, although some parts of the Bathurst Estate and Cirencester Park are open to the public, the inside of this curved hedge is not. However, from the road outside (Park Street) you still get an impressive view and can appreciate the scale of this wonderful hedge as it towers high above the boundary wall. You can also climb the tower of St John Baptist Church on selected dates over the summer months (fee applies), for a bird's eye view of the colossal hedge.

Rising to a height of 12 m (40 ft), this closely clipped curving hedge has presented its pristine outer face to the town of Cirencester for the past 300 years

Benington Yew

7

Location. In the churchyard of St Peter's Church, Church Green, Benington, Hertfordshire SG2 7LH. **OS grid ref.** TL297235

Lying down on the job

Far longer than it is tall, this male yew looks more like an overgrown laid hedge on a field boundary than a tree. Situated south-west of the church, it fell to the ground long ago – not suddenly uprooted by wind and weather, but slowly toppling earthward in a gradual process, to which its twisting, strained trunk base bears testament.

As it neared the ground, the tree partially propped itself up on one of its branches. It has also been give extra support, as a couple of sturdy wooden props have been added along its length.

The first 1–2 m (3–6 ft) of the trunk are completely hollow with lots of dead wood, but also plenty of living wood supplying vitality along its length, ensuring the yew's survival. In the hollow, the tree has put down aerial roots to bring extra sustenance to the growing areas. There is a small amount of growth in this initial section of trunk, but it is the vertical twisting limbs further along that give a clue as to how long the yew has been in this recumbent position – maybe 80–100 years, judging by their girth.

The church and churchyard in the small picturesque village of Benington have been in existence in one form or another from the 9th century. Adjacent to the church are the ruins of a Norman keep, built in 1138 and demolished by Henry II in 1177.

Despite toppling over long ago, this yew is still thriving, with several branches now growing vertically to form a row of new trunks

Bettws Newydd Yews

8

Location. In the churchyard of St Aeddan's Church, Bettws Newydd, Monmouthshire NP15 1EQ. **OS grid ref.** SO362058

A Welsh trinity

Meaning a place of prayer or an oratory, 'Betws' occurs in place names all over Wales. In the small village of Bettws Newydd (New Oratory), St Aeddan's churchyard is home to three large old yews – all male and all very different from each other – as well as a small, largely 15th-century church, which has features from the 12th and 13th centuries.

The oldest yew has put down a now very substantial aerial root, into its hollow centre, that is keeping the tree alive and upright

On the west side of the church is the largest yew – an exceptional ancient tree that is far bigger than the other two. Its huge trunk is all but a shell, mostly now dead wood and barkless, giving it a pale weather-beaten appearance. You might wonder how such a fragile-looking trunk could hold up a canopy of this size, but in the centre of this hollow giant is a large aerial root, with a 2 m (6 ft) girth, itself many years old, giving support and new life to the tree.

The following rhyme about this ancient yew is carved on a piece of wood in the church porch:

The star which shone on Bethlehem also shone on this noble tree –
It stood here still when Norman Knights claimed their victory –
As King Charles lost his royal head this tree made growth anew –
Thanks be to God that we still have our mighty Bettws Yew.

Close to the churchyard boundary wall, to the north-west of the church, stands the second of the three yews. It has a straight, fluted trunk that is clear of branches for the first 3 m (10 ft). The trunk looks quite sound, but around the back there is evidence of the bracket fungus *Laetiporus sulphureus* (chicken-of-the-woods), which causes brown rot of the heartwood. So this yew is definitely in the process of hollowing. Some of its lower branches have dipped down to the ground and have layered themselves, taking root, both supporting the tree and producing new clones.

The final tree is around the back of the church, and slightly bigger than the second, with a large flared base, partly covered in ivy. It is already quite hollow and is filled with a mass of internal, twisting aerial roots.

Bleeding Yew (Nevern)

9

Location. **In the churchyard of St Brynach's Church, Nevern, Newport, Pembrokeshire SA42 0NF. OS grid ref. SN083400**

Not for the squeamish

This yew has inspired lots of local stories, myths and folklore, as you can imagine with a tree that appears to be oozing red blood! Many trees and shrubs will bleed sap quite profusely if cut, particularly in spring when the sap is rising. Birches and maples are prime examples, and in spring the sugary rising sap of some maple species can be collected and turned into maple syrup.

However, one of the yews in St Brynach's churchyard is particularly remarkable, because it continually exudes red sticky sap, and has done so for as long as anyone can remember.

Another curiosity is that it bleeds from a strange place. Sap usually rises around the outer edge of a tree, in the living tissue just under the bark. The heartwood in the central part of the trunk and branches is dead structural wood and shouldn't bleed. But this is exactly where the sticky red ooze is coming from. The cause of this odd phenomenon remains unexplained.

One of the many myths about this tree is that it bleeds for the 'wrongful hanging of a monk, for a crime he didn't commit'. Another is that it bleeds out of sympathy for Jesus, crucified on the cross. It is also said that it will continue to bleed until a Welshman once again sits on the throne at nearby Nevern Castle.

Entering through the churchyard gate, the Bleeding Yew is the second yew on the right, in an avenue of yews that leads to the church porch, and the bleeding takes place on the far side of the tree.

After visiting the tree, take a look too at the wonderful Vitalianus Stone, to the east of the church porch, which is inscribed with 5th century Ogham script (an early medieval alphabet from Ireland). Also of note in the churchyard is an ornately decorated Celtic high cross dating from the 10th century, while set in the windowsill of the church's nave is a carved braided cross also thought to be from the 10th century.

It's a mysterious phenomenon – for years, sticky red 'blood' has dripped from the heartwood of the yew in St Brynach's churchyard, but no one can explain why

Borrowdale Yews

10

Location. Near Seathwaite, Cumbria CA12 5XJ. **OS grid ref.** NY235121

Poetic inspiration

Towards the southern end of the Borrowdale valley, close to Seathwaite, on the hillside above the River Derwent, is a group of trees that are the most famous yews in England, immortalised by the poet William Wordsworth. He was so inspired by this small group of trees that he wrote the poem *Yew Trees* about them in 1803. In the poem, he describes them as the 'fraternal four'.

> *...But worthier still of note*
> *Are those fraternal Four of Borrowdale,*
> *Joined in one solemn and capacious grove...*

Fraternal implies that the trees are male, but they are in fact all female (at least, the three still standing are). The four became three in 1866, after a storm felled one of them. The gender of that tree is unknown.

These ancient yews are now protected by a short stock fence, creating a compound around them, to prevent herbivores damaging them.

As you drive along the road to Seathwaite from the village of Seatoller, you will see how well wooded the sides of the valley are. These are some of the best woodlands in the Lake District, particularly the upland oak woods, which are remnants of a wider temperate rainforest that covered the whole Atlantic seaboard of western Britain. Among these oaks, even during spring and summer, the dark green foliage of the Borrowdale Yews stands out clearly.

Climbing up the hill to the enclosure, you first come across a National Trust interpretation panel with a timeline tracing the possible life of the yews. Once over the stile and inside the enclosure, the first yew you come to is a tall tree with a fairly open canopy and the narrowest trunk of the trio. DNA analysis has shown that this tree has the same DNA as the yew closest to it, which has the widest trunk of the group. My hunch is that the first tree is quite likely to be an old rooted branch, or layer, from the broader-trunked yew growing just above it. This would account for them sharing the same DNA.

This first yew once had a double trunk, and the remains of the part that broke away still lie beneath it.

A fallen branch from the second yew has been dated at 1,500 years old, although the tree itself may well be even more ancient

It also has quite a bad lean downhill, but has grounded its lower branches on the downward side, supporting and stabilising itself. This lean was probably caused in part by the canopy of the second yew above it, before that tree lost a substantial limb during a storm in 1998.

The second tree has been reduced to a tall, hollow shell, although it is very much still alive. The loss of the large limb in the 1998 storm gave dendrologists the opportunity to do a ring count and get an idea of the age of this particular tree. A sample cross-section of the fallen limb, sent to Newcastle University, was dated at 1,500 years old, and the tree itself could be even older. In 2005, this tree suffered further damage, when a powerful storm blew out the entire canopy, much of which still lies on the ground around it. Since then, the tree has started to grow a new canopy from the living sections of trunk (as you can see in the photograph), which can clearly be distinguished from the dead, bleached white wood. The hollow trunk is filled with decaying matter, and there are almost as many ferns growing on the inside as there are on the outside.

The last of the three standing yews is probably in the most delicate condition, completely hollow, with around half of its trunk being dead wood. Fortunately, the living section has grounded branches, giving it some stability. Considering its poor condition, it has a fairly large, healthy-looking canopy. There is a lot of stone rubble on either side of the trunk, suggesting that it might have once formed part of a boundary wall.

The smallest of the three, this yew is a clone of the one growing slightly above it, so probably resulted from a grounded branch that rooted long ago.

The last of the 'fraternal four' was completely uprooted during the Great Storm of 1884. A dead section of trunk lying just outside the enclosure is believed to be the fallen remnants of this tree.

Right The third yew is the most fragile, being totally hollow, but it gets extra support from having several branches touching the ground

Left The smallest of the three, this yew is a clone of the one growing slightly above it, so probably resulted from a grounded branch that rooted long ago

Directions From the Honister Pass, on the B5289, take the lane signposted to Seathwaite. There is some limited parking along the lane, but it's easier to park in Seathwaite Farm, where there is an honesty box to pay for parking. Take the small wooden gated bridge over the river, turn immediate right and follow the track along the river for about 700 m (half a mile). The yew trees in the enclosure are easily seen above on the left.

11

Breamore Yew

Location. In the churchyard of St Mary's Church, Breamore, near Fordingbridge, Hampshire SP6 2DF. OS grid ref. SU153188

A new lease of life

This tree is very different to most of the other ancient yews you'll come across, whose hollow trunks tend to become fragmented over time. Instead of being made up of several old trunk fragments, this tree appears to be composed of about ten young trunks. These grow in a circle, in and around where the original trunk would have stood, and the raised mound in the centre is almost certainly its decayed remains. Considering the many pieces that make up this unusual yew, its canopy is large, spreading and healthy.

It's difficult to work out exactly what happened in the past to this intriguing tree. There is an account of the 'complete destruction of the head' in 1958 in *The Yew Trees of England* by E. W. Swanton. This might suggest that the young trunks are suckers that arose around the stump when the tree's head was destroyed, but that is not the case.

In 1897, this tree is mentioned in *The Yew-Trees of Great Britain and Ireland* by John Lowe. It gives the trunk girth as 9 m (30 ft) and notes that there are '8–10 young trunks, a foot or more in diameter'. So most of the trees you see today are probably the outer, remaining fragments of what was once a huge trunk, particularly those that show signs of decay on their inner sides.

Those that don't could be old aerial roots laid down before the main trunk decayed. One of these in particular is growing inside the outer circle, with its root system spreading over and into the decayed matter in the centre.

Most of the fragment trunks have their main growth on their outer sides, and are all female. But there is also a single male trunk growing close by, with branches on all sides. This is probably a self-sown seedling, but it may have been planted there deliberately to succeed what was considered to be a dead or dying tree.

In a gap in the circle of trunks, a group of stones is visible – these were perhaps used to block up a hole or cavity long ago. On the eastern side of the tree are some stone coffins, which were brought here in 1898 from

excavations at an Augustinian priory less than a mile away on the west bank of the River Avon.

This yew stands on the south side of a beautiful large Saxon church. Much of the original building from the late 10th or early 11th century still survives today and the church is considered to be one of the most important buildings of Saxon origin in southern England.

The mound inside the circle of outer trunks is the decayed remains of the original tree

Claverley Yew

12

Location. In the churchyard of All Saints' Church, Church Street, Claverley, Shropshire WV5 7DS. **OS grid ref.** SO793934

Two sides to the story

Certainly one of the oldest trees in Shropshire, this ancient yew can be found to the north east of All Saints' Church and is definitely a tree of two halves. The tree's outer shell is just that, a shell, composed mainly of dead sapwood, gnarly and covered in raised areas where lots of epicormic shoots would have once hidden its trunk. This now resembles dried-out and silvered sea coral. Old yews often produce leafy epicormic shoots from dormant buds under the bark when they are under stress or when their crown is dying back.

Viewed from one side, this tree is mostly made up of a large area of the dead shell. But on the other side you will find plenty of healthy growth, mainly coming from one small section of trunk that has many strong branches bearing lots of healthy foliage.

If you look up into the tree you will see some wonderful old ironwork (probably Victorian) – a chain with long flat rectangular links and an equally antiquated turnbuckle tensioning device. These would have held sections and limbs of the tree together in times past.

The inside of the church is pretty amazing too. There is a rough-hewn bowl of a Saxon font from the 7th century, while wonderful medieval wall paintings adorn the nave above the arcade. This equestrian frieze is thought to represent the seven Christian virtues and the seven pagan vices, and dates to around the 13th century. It was rediscovered under coats of whitewash during restoration work in 1902.

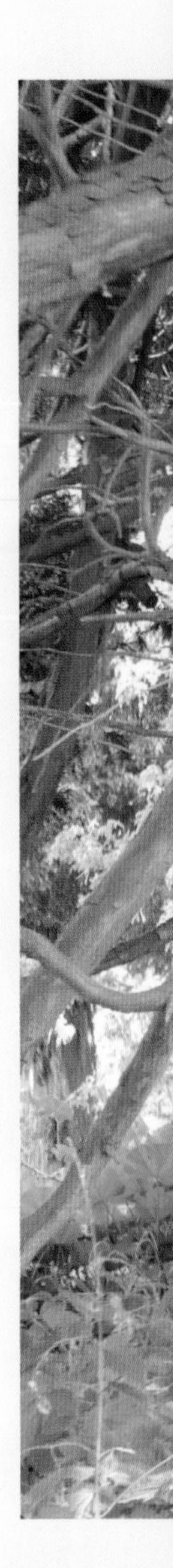

Half of the tree's vast trunk is a dead shell, while the other half has plenty of healthy growth

13

Coldwaltham Yew

Location. **In the churchyard of St Giles' Church, Church Lane, Coldwaltham, West Sussex RH20 1LW. OS grid ref. TQ023165**

An architectural gem

This tree reminds me of some of Antoni Gaudi's wonderfully creative buildings in Barcelona. It is a female yew with a completely hollow trunk, but unlike most other ancient hollow yews, which usually open from ground level, this one retains its outer shell all the way around its base. The major openings are at around waist height, allowing you to peer easily into its cavernous interior.

Hollow this yew may be, but it still manages to maintain three large vertical, fluted limbs, with much branching higher up and some large, twisting lower limbs. Although the tree is fairly sparse, with its foliage mostly towards to top of its canopy, it still makes a truly massive statement in the north-west corner of this tranquil churchyard.

There is a small metal plaque close to the tree that suggests it is one of the 12 oldest in England. The church of St Giles itself is in parts around 800 years old and is thought to have been preceded by an earlier Anglo-Saxon chapel on the same site, whose Saxon font can be seen in today's church. The magnificent old yew probably pre-dates both of these buildings.

During the 1960s, the tree was given a Tree Preservation Order (TPO) to save it from possibly being cut back or worse, after land adjacent to the churchyard and the tree was given planning permission for two building plots. The TPO does provide a measure of protection, but as with most of the old trees in the UK, it is the locals and tree societies who actively preserve it.

With its many holes, ridges and towering pinnacles, this tree does bear an uncanny resemblance to Gaudi's Sagrada Familia basilica in Barcelona

14

Crom Castle Yews

Location. National Trust Visitor Centre, Crom, Upper Lough Erne, Newtownbutler, County Fermanagh BT92 8HR. **OS grid ref.** SA454877

In a loving embrace

On the shores of Lower Lough Erne, close to the original entrance and ruins of the Old Castle at Crom, you can see yew branches that from a distance look like the huge spreading canopy of a single tree. In fact, these branches are from two very characterful trees, said to be the oldest yews in Ireland. It is claimed that they could be up to 800 years of age.

Walking into this huge canopy is like entering a dark, wooded vault. Under here you find two mature yew trunks growing side by side, with their branches intermingling, joining the two crowns as one. One of this pair is a male and the other a female, both with amazingly sinuously twisting branches and growing on mounded areas higher than the surrounding ground outside their canopy, which has a diameter of around 35 m (115 ft).

The female tree is quite a bit larger than the male, and probably older (as pictures exist showing it as a single tree with a vast propped canopy). It has dozens of low twisting branches coming down to the ground, creating dark cave-like spaces, with areas under these where the ground is well worn from the many children and adults crawling around in this almost other-worldly space. I was really amazed by how the branches twist and spiral, and even in the upper canopy, the limbs twist and turn all over the place.

The smaller male tree is in fact the taller of the two. It has fewer low twisting branches, and some grow directly upwards from low down. Many of the branches that once grew between the two trees have been cut back, separating them at a low level.

The earliest known reference to the trees is from 1739, when they were described as having their branches trained horizontally, clipped and propped with wooden posts near their ends. This produced a large umbrella-like canopy, under which parties for up to 200 guests were once held!

These two interwoven yews are the only trees in Northern Ireland to feature in the list of '50 Great British Trees', compiled in 2002 in honour of the Queen's Golden Jubilee

15

Crowhurst Yew (Surrey)

Location. In the churchyard of St George's Church, Crowhurst Lane, Crowhurst, Surrey RH7 6LR. OS grid ref. TQ390474

Don't forget to knock

The largest yew in the county of Surrey, this remarkable tree has many stories associated with it. It stands on the west side of the 12th-century church, and apart from its impressive size and girth has two main claims to fame.

First, it is noteworthy for its great age – various sources suggest it may be up to 4,000 years old, but this is purely speculative. Nevertheless, it surely pre-dates the 12th-century church that now stands alongside, so must somewhere near 1,000 years old at least. The first recorded measurement of its girth was in 1630, when the trunk had a circumference of over 9 m (30 ft). The tree's other distinguishing feature is the wooden door in one of the openings in its huge trunk. This rather lopsided door (more of a gate really) does not hang or open as well as it probably once did when it was first installed in the early 19th century (probably in the 1820s). At that time, the tree was already hollow at its base, but was manually hollowed out even more to make it large enough to fit a small round table with a wooden bench.

When this work was carried out on the tree, a cannon ball, assumed to be from the English Civil War in the 17th century, was reportedly found imbedded in the trunk, and the wound had already begun to partially callus over.

The hollowed-out room also apparently had a roof, although it is not clear whether this was added or whether it was a natural part of the tree. Unfortunately, a large part of the tree's crown was damaged during a storm in 1845, which was perhaps when the roof fell in.

Now only the door remains, and the trunk has a few more openings due to continuing decay – not just an extra back entrance but a side one too! So it is no longer the enclosed internal space it once was, and is a lot more draughty. Still, standing inside this hollow giant is like being in a grotto. With its twisted and gnarled inner surface, you can only wonder at what it would have been like to be seated inside for a meeting or party.

Around 200 years ago, the yew's massive hollow trunk was made into a room, with a door, a roof and a table – up to 12 people could reportedly be seated inside

Wooden props have also been added over the years to help support some of the tree's limbs, but for all this, the tree still looks to be fairly healthy. Because of its bizarre and incongruous door, this is one of the most photogenic trees – it looks like it could have come straight out of a fantasy story.

Crowhurst Yew (Sussex)

16

Location. In the churchyard of St George's Church, Crowhurst, East Sussex TN33 9AJ. OS grid ref. TQ757123

Flares are back in fashion

This tree is often confused in the literature with an ancient yew in Surrey (see p 72), as they both grow in villages called Crowhurst and in the churchyards of parish churches dedicated to St George. They are also similar in size, but very different in appearance.

This yew is one of my favourite trees. When we plant trees today, we're very exact about planting depth, and the trunk of a well-planted tree should show good basal flare from a fairly young age. This remarkable tree looks to me, although scaled up many times in size and age, like a well-planted tree with really good basal flare!

This ancient female yew is protected by a circular fence of metal railings, erected in 1907 by Col. P. R. Papillon for its protection, and is the largest of three yews growing in this churchyard. Its trunk is hollow but has some good internal stems.

The side of the tree facing the church has almost two distinct parts, the larger and more upright section has an opening that is edged with almost white dead wood, giving it, I think, a rather appealing appearance. The limbs of the tree are being allowed to escape their cage and grow over and beyond their captive circle, touching the ground in an attempt to naturally prop and support the old timer. Eventually, if allowed, they will root into the soil to produce new youngsters to continue the line. There is a leaning part of the trunk on the far side that is slowly moving away from the main tree and will very likely come to rest on the railings at some future time.

Although the crown has some cabling supporting the upper branches, and is generally thin, the tree looks to be in fairly good health.

The other two large yews in the churchyard are of a similar size to each other, but quite a bit smaller than the caged yew. They are also both female and thought to be approximately 600 years old, planted around the time the current church was built.

The plaque on the railings claims that this mighty yew dates from at least 1066 – when the Battle of Hastings was fought and King Harold was the owner of the Manor of Crowhurst

1066 – 2016
This ancient yew was here in 1066
when King Harold owned the
Manor of Crowhurst.
In this 950th anniversary year of
the Battle of Hastings we remember
his close links to this part of Sussex.

17

Cudham Yews

Location. In the churchyard of St Peter and St Paul's Church, Church Approach, Cudham, Kent TN14 7QF. **OS grid ref.** TQ445599

An enchanting couple

They say that beauty is in the eye of the beholder. Well, in my eyes at least, these two old yews, which probably pre-date the 11th-century church, have a beauty that is unique to ancient trees. The two yews stand on the south side of the early Norman church, on either side of the porch – the male tree at the western end and the larger female at the eastern end.

The male yew is slightly smaller in girth when both are measured low down at the same point, but its trunk bulges out to almost the same girth at 1 m (3 ft) from the ground. The tree's three main limbs spread out like a giant trident. The largest limb is the central one, which towers up, adding great height to the tree, and is hollowing at its base. The trunk also has an area that is partially hollow.

The female tree is the larger of the two and has a huge opening from top to bottom on its north side. A small piece of metal railing has been added some time ago, partially closing off this opening, presumably to keep people out. This railing is now being swallowed into the tree by new wood being laid down as callus. On the side opposite the opening, the rippled, gnarly old trunk looks like a tree that has been badly put back together! Unfortunately, as with so many magnificent yews, there is evidence of old fire damage to the hollow interior of the trunk.

The lovely little flint church sits on a local high spot at the edge of the village. The earliest parts of the current church are from the Norman period, but according to the church guide a Saxon script has been found that makes reference to a church being here in AD 982.

Left The female tree is starting to swallow the small section of metal railing that blocks off the opening

Right The male yew, partly hollow and with deeply fissured bark, is full of character and natural beauty

18

Cyffylliog Yews

Location. **In the churchyard of St Mary's Church, Cyffylliog, Ruthin, Denbighshire LL15 2DW. OS grid ref. SJ059578**

Setting clear boundaries

The five old ladies of Cyffylliog stand guard around the boundary of the churchyard, which is thought to have been a site of Christian worship since the late 12th century

Standing guard on the churchyard boundary, on all sides of the church, are five old female yews. When you enter the churchyard through a side gate to the north-west of the church, you immediately encounter the first of the trees and the middle one in size. It has two large trunks that are growing away from each other, making a large V-shape. Both trunks are partially swathed in ivy. The tree is growing on a mound encased in local stone, so it sits much higher than the path.

Close to the church porch is the second largest of the trees, which, like all the yews in this churchyard, is growing on a raised area. The ground at the base of this yew is almost level with the top of the stone boundary wall, around 1.5 m (5 ft) high.

The most impressive of the five, and by far the biggest, is a colossal yew, again standing on a mound, which makes it look even more majestic. The mound is more than 1 m (3 ft) higher than the surrounding ground. Growing on the east side of the church, this is the only one of the five that is not actually on the churchyard's boundary edge. But maybe it once was?

The crown of this tree is immense and has a spread of more than 22 m (72 ft), with two huge limbs heading skyward. These are multi-branched, adding to the tree's stature. Lower down, the limbs grow out at a 45-degree angle, creating a wide spreading canopy that overshadows the gravestones and tombs that surround it. This old yew shows little sign of hollowing.

The last two yews are the smallest. The most significant of these has a very tall, single, fluted trunk and grows close to the churchyard's main entrance, on the boundary and against the old stone hearse house, which stands outside the churchyard.

Like many of the churchyards I have visited, this one is sited on an area of high ground, with the fields to the east of the boundary wall being at least 2.5 m (8 ft) lower than the churchyard. These fields run down to a small stream, less than 100 m (330 ft) away in the lower part of the village – another feature frequently found near churches with old yews.

Darley Dale Yew

19

Location. In the churchyard of St Helen's Church, Church Road, Darley Dale, Derbyshire DE4 2GL. **OS grid ref.** SK266629

Plea for protection

Right Reputedly over 2,000 years old, this yew is of great age as well as great corpulence – its trunk circumference of over 10 m (33 ft) is among the largest in the UK

Dominating the entrance of the porch on the south side of the parish church of St Helens is a female yew with an absolutely huge trunk, well over 10 m (33 ft) in circumference, and full of deep nooks and crannies. Everything about this tree is enormous. The two main large vertical limbs are so big that they would individually make impressive trees in their own right.

The tree is surrounded by a Victorian metal railed fence, with decorative trident-like spiked tips. The trunk is covered in a ring of burrs at ground level, becoming deeply fluted up to the first branches. Its lower branches, like those of many old yews, are trying to reach the ground, dipping down over the fence as if to escape their confinement. Rather incongruously, a group of stone tablets commemorating famous battles of the Second World War sits at the base of the tree.

The tree itself has suffered some damage over the centuries – not just caused by storms, but by visitors too. In 1863, The Times published a letter, purportedly from the 'Old Yew Tree, Darley Churchyard', making a desperate plea for visitors to be more respectful and not 'cut, break, and mutilate my poor old person in all conceivable ways' or 'immortalize themselves by cutting their names all over my bark'. One visitor apparently even 'drew out a saw, and actually set to work to cut out a great slice of my very flesh, which, but for the lucky intervention of the clerk, he would soon have accomplished'. The letter went on to claim: 'I am no common tree... My age is fabulous, and learned naturalists now calculate that I must have been born 300 years before the gospel was planted in this country'. An incredible and resilient tree indeed – now well protected behind its spiked railings!

The church is very interesting too – the original building is thought to have been founded towards the end of the first century, and some Saxon and Norman workmanship remains.

Below Vintage chromotype postcard showing the old yew tree

An ancient stone carving of Sheela-na-Gig, a female figure linked with fertility and creation, can be seen on the keystone of the archway inside the church. And in the south entrance porch, groups of wonderful stone coffin lids, which appear to be of Saxon origin, stand propped each side.

20

Defynnog Yews

Location. In the churchyard of 's St Cynog's Church, Defynnog, Powys LD3 8SD.
OS grid ref. SN925279

The jury's out...

Four ancient yews (or perhaps just three – it is open to debate) sit in the roughly oval-shaped churchyard of St Cynog's on the edge of the small village of Defynnog. It is claimed that one of them is the oldest tree in Europe.

The Defynnog yews surround the church. The second largest of them grows on the south side of the building and is many branched, with an abundance of bristly growth on the trunk. The tree on the north-east side sits within a walled mound. It is hollow, with a very sinuous, tangled look to the main trunk, and has lots of small branches all the way round, many coming down towards or actually touching the ground.

The largest of the trees in the churchyard is on the north side of the church and recently has been the topic of much debate. Some believe that both it and the smaller yew, just 5 m (16 ft) to the west of the larger tree, are one and the same. DNA tests have shown that they are genetically the same, but this doesn't automatically mean that they were once a single tree. It could be that the smaller tree is a layer (or rooted branch) from the larger one, which would account for them being genetically identical.

I can see both sides of the argument. Even though they do look like separate trees, and not like two outer shells of a fragmented tree, they do have open sides facing each other. I think that these could be the growths from aerial roots left from now long decayed fragments. This would have made the original tree absolutely huge, and just because we have no evidence of trees of that scale, it doesn't mean they didn't exist!

Even if the larger tree is a separate individual, what a tree it is! It has a huge circumference, with nine large limbs that are covered in burrs and leafy epicormic growth arising from directly below the bark. There is a central platform-like area filled with small roots, around 2 m (6 ft) above the base of the tree, with the yew's large limbs radiating around it.

Is this vast yew really the oldest tree in Europe? For me the jury is still out...

As for the church, it dates mainly from the 15th century, but parts of the earlier 11th-century building still exist in the stonework of the north vestry wall. In the porch is a long pillar-shaped 5th-century Roman tombstone, along the side of which there is also some Ogham script (an early medieval alphabet from Ireland, but also found in Wales). At the top is a Celtic ring-cross, thought to have been added sometime between the 7th and 9th centuries. This Roman tombstone was found in the wall of the church tower, indicating that this site has ancient history.

The largest yew is producing lots of epicormic growth around its trunk – fresh green foliage sprouting directly from under the bark

Didcot Yew

21

Location. In the churchyard of All Saints' Church, Lydalls Road, Didcot, Oxfordshire OX11 7EA. **OS grid ref.** SU519905

Phoenix rising

This is a very impressive ancient male yew, not just because of its age and stature, but also because it has survived more than two separate arson attacks. The most recent was in 2005, and caused severe damage to the tree's canopy.

Standing to the south-west of the church, this male yew has a large trunk that flares widely towards its base, with dark red, heavily flaking bark running in wavy, parallel lines up the trunk, interspersed with bands of lighter, silvered wood.

There are two openings into the trunk – a small one on the side closest to the church and a much larger one on the opposite side. The latter comprises several smaller openings, but you can still easily access the hollow interior. Looking into this space, it's very easy to see the arson damage. The whole of the inside is completely charred, and unfortunately seems to have become a dumping ground for discarded cans and bottles. Much of the canopy was destroyed by the fire too, or removed afterwards to make the tree safe.

Despite this, two of the main branches, which mostly survived the fire, are flourishing, and the tree has put on good growth. It is now beginning to replace the once much larger canopy – a photograph from the early 1900s, in the book *Didcot Through Time* by Brian Lingham, shows the tree standing almost as tall as the church spire. The new growth, because of the lack of branches, has become a bit lop-sided, but will hopefully continue to thrive and eventually reach the height of the church spire once again.

All Saints' Church dates back to the 12th century, although much of the current building is from the 14th century, with later Victorian renovations.

Fire damage has resulted in a very lop-sided tree, but the remaining canopy seems very healthy

Discoed Yews

Location. In the churchyard of St Michael's Church, Discoed, Powys LD8 2NW.
OS grid ref. SO276647

Celtic couple

Originally circular in shape, the small churchyard of St Michael's is now just a semi-circle, within which stand two ancient yews. The church itself is medieval in parts, but was extensively restored in the 19th century. Just outside the churchyard entrance is a small well, built around a spring. The shape of the original churchyard and the presence of a spring, which may once have been within the churchyard, are characteristic of Celtic religious sites, so perhaps this spot was in use long before the current church was built.

The first tree you come to as you enter the churchyard is the smaller of the two yews (but still a huge tree). This is a female yew and sits slightly raised up on the south-west boundary, above a small entrance driveway. It has many low branches, making the huge fragmented trunk initially difficult to see. Ducking under these reveals the true scale of this hollow giant. This tree can only be viewed easily from one side, as the other side is in a private garden, which presumably would have once been part of the original churchyard.

Below This ancient fragmented yew dominates the north side of the church

Above The larger tree, which has split into three sections, was potentially already several thousand years old when the church was built

The second tree you come to is the larger of the two, a male yew that dominates the north side of the church, and its trunk has an exceptional girth of over 11 m (36 ft). But this is completely fragmented into three large sections. One of these has bark that is much lighter than the other two – it is almost silver and looks nearly dead, although it is very much alive.

Both trees have fairly full crowns with good leaf coverage and appear to be in robust health.

Local poet Edward Storey has found great inspiration in the Discoed Yews, and in the following composition he considers the great age of the larger tree.

Beyond Tree Rings

How could this tree's biography ever be written? Had sight reached beyond boundaries, it might have watched the Hebrews' flight from slavery long before David's psalms were sung. Closer to where its roots were struck in Wales, it saw the earliest shepherds come with weavers and harpists, their eyes bewildered by the first conjuring tricks of spring. Now it looks on another world of slaves which may have measured space but failed to lengthen hours by inch – our hill a stone we've briefly settled on. And neither it nor any dying star can say where the next birth-seed will be sown. Even the sun may forget how to make shadows, leaving only silence beyond tree rings . Yet why should we stop singing? We too are freed by music that breaks chains and, for the split second of a millennium, join in the dance to blossom, root and stone.

Doveridge Yew

Location. In the churchyard of St Cuthbert's Church, Church Lane, Doveridge, Derbyshire DE6 5NP. **OS grid ref.** SK114341

Triumphantly arched

As you walk through the churchyard gate, you find two relatively young yews standing like a pair of sentries guarding a tunnel through the dense low canopy of a large ancient yew. The dropping branches are held above the path by a wooden framework, creating a sheltered tunnel that leads up to the church entrance. The church itself dates from the late 12th century, although it must have replaced an older building, as the *Domesday Book* of 1086 mentions a church in the village of Doveridge.

The ancient female yew from which the 12 m (40 ft) long tunnel of branches is formed looks at first glance to have a complete and intact trunk. But if you walk around it, you'll see it is in fact mostly a shell, completely hollow and with an opening on one side that is about a third of its total circumference. The trunk that remains has very thin walls.

There are also many branches twisting and turning away from the main tree, looking like a many-tentacled monster when you stand in the gloom beneath the canopy.

Below The tree's vast canopy, with its drooping limbs, spreads right across the path, creating a tunnel-like effect

Right An old chain encircles the upper trunk, keeping the massive weight of the branches from splitting the tree apart

The upper part of the tree was once held together by two large old chains, the higher of the two having snapped long ago. The lower chain is still in one piece, holding the trunk together, but it has not stopped one of the mighty limbs dipping right down to the ground under its own weight and brushing against the church wall.

On the opposite side of the canopy to the tunnel, there are more wooden props under many of the branches, attempting to keep them above some of the gravestones that surround it. But the tree is determined to do what comes naturally, and that is to put its branches down on the ground for support, which will hopefully root to make a new generation of clonal offspring.

The legendary Robin Hood, so local folklore has it, was married under this very tree, but not apparently to Maid Marian! According to an old ballad, his bride was Clorinda, queen of the shepherdesses.

Downne Yew

24

Location. In the churchyard of St Mary the Virgin Church, High Street, Downe, Kent BR6 7UT. **OS grid ref.** TQ432616

Survival of the fittest

The fragmented state of many ancient yews, caused by extreme old age, only adds to their mystical air, and this venerable female yew is a prime example. The trunk is now completely hollow and divided into two main sections. Its gnarly old frame, full of nooks and crannies and smaller holes, invites you to peer in and take a closer look. The cavernous hollow, open on both sides, comes together again at around 2 m (6 ft) high. From here, the main branches fan out in a tangle of growth, producing a canopy that spreads above and over the perimeter wall of the churchyard on one side and that almost touches the church porch on the other. Despite its hollow interior, the tree is in remarkably good condition.

Charles Darwin undoubtedly visited this churchyard and admired this beautiful ancient yew

Charles Darwin, the 19th-century naturalist best known for his work on the science of evolution, lived with his family in Downe House, just a few miles from the church. His brother Erasmus Darwin was buried in this churchyard during Charles's life time. Surely the great naturalist must have also been in awe of this hollow giant of a tree, which would have been every bit as impressive in his day as it is now. Some of Darwin's children are also buried in the churchyard, to the west of the path. Darwin himself is interred at Westminster Abbey, but a sundial on the south wall of the tower is dedicated as a memorial to him.

This impressive yew is almost certainly older than the medieval church it stands beside. But this building was preceded by a chapel built at the end of the 13th century, under the instruction of Prior Henry of Christchurch. The current church has been much repaired over the centuries – there are records of rebuilding works in the 15th century, and more recently towards the end of the Second World War, after the church was damaged by a flying bomb in 1944.

Druids Grove Yews

25

Location. Druids Grove, Norbury Park, near Mickleham, Surrey, RH5 6ET.
OS grid ref. TQ158531

An exceptional cast of characters

Druids Grove is a group of well-hidden old yews, whose rich diversity of forms makes this a truly exceptional and special place. Most of the yews have short, stout trunks that break around 2–3 m (6–10 ft) high into several branches that shoot up many metres, heading for the light above the woodland's dark canopy. Some of the yews, however, have their branches spreading down and around them (as is often the case with wild yews), creating tunnels and looking very mysterious. One tree is quite different from the rest, having a taller, more slender trunk, with fairly even branching. I guesstimate this tree to be 18–20 m (60–65 ft) tall. One of the first of the large yews you see as you enter the grove from the south is on the right of the path with a solid-looking trunk. Soon after, you come to another large yew, also on the right, but this one has a big, bulging, hollow trunk with a large aerial root inside the cavity.

The path through the grove is fairly level, but the ground on either side rises and falls steeply. The grove itself sits in the middle of woodland in Norbury Park, high above the River Mole, on an eastern slope facing across the valley towards Box Hill. There are lots of yews of all ages within the woodland, mixed in with mainly ash and sycamores, plus a few beeches. Designated a Site of Special Scientific Interest (SSSI), the woodland and surrounding area are owned by Surrey County Council and leased and managed by the Surrey Wildlife Trust.

Druids Grove first appeared on an Ordnance Survey map in 1873, and is also known as Druids Walk. The Anglo-Saxon king Edward the Confessor owned the Manor of Norbury during the 11th century, and Norbury Park and Manor are mentioned in the *Domesday Book* of 1086.

Some of the yews in Druids Grove have grown tall, reaching for the light above the surrounding dense canopy of foliage

In more recent times, the Great Storm of October 1987 took a heavy toll on this woodland, with around half of its mature yews, and many of its mature beech trees, uprooted or severely damaged. Along the path to the grove there is still much evidence of this. Huge beech trunks, without their canopies, stand like monoliths, bearing testament to the power of nature.

Other trees lie where they fell 30 years ago, their fan-like root systems clearly visible, showing just how amazingly shallow their hold on the hillside was. Within the grove, I measured a fallen yew lying alongside the path at over 17 m (55 ft) in length. Many of the fallen yews, though, have continued to grow (as yews do), sending up new vertical stems from their now horizontal trunks.

Directions To get to the oldest trees in Druids Grove, there is a walk of about 30 minutes, which can be slightly challenging in parts. This is particularly so when the paths are wet, as the chalky soil can be very slippery, especially in the steeper sections. There is limited parking at the end of Cowslip Lane (off the A24 near Mickleham). From here, follow the wooden waymarker's directions for Leatherhead, along an initially concrete path. Continue over a small bridge crossing the River Mole, turn left along a field and into the lower part of the woodland. Then turn right and head north, and after a short distance you will come to signs for Druids Grove.

As with many woodland yews, their lower branches spread down around themselves

Dunsfold Yew

26

Location. In the churchyard of St Mary and All Saints Church, Church Green, Dunsfold, Surrey GU8 4LT. **OS grid ref.** SU998363

Hole-hearted approval

This short, stocky male yew stands close to the south porch of the church. Its trunk is completely hollow and fragmented at the base, but with some parts joined higher up at around 2 m (6 ft). The largest opening is on the side opposite the porch, with other openings on all sides.

The trunk is in three main sections, with its lower limbs stretching out at right angles, some now dead. Many of the lower limbs, though, have been removed and the wounds painted with bitumen (an old method of protecting pruning cuts), but these show no signs of callousing. There are accounts that many of these branches were supported on wooden poles to keep them above the gravestones, and one was almost leaning on to the roof of the church porch.

Inside the hollow trunk are some areas that contain concrete – it was common practice in times past to fill voids in this way. There are also historical scars on the upper limbs from old bracing chains that would have been used to support some of the weight when it was a larger tree, to stop it pulling itself apart.

Not far from this large yew, on the west side of the church, is another interesting, though much smaller and younger male yew. It was planted in 1977 to celebrate 25 years of the Queen's reign, as its plaque affirms: 'A yew tree has stood near the door of Dunsfold church for 700 years. This tablet marks a yew planted to commemorate the silver jubilee of the accession of Queen Elizabeth 2nd'.

In the churchyard there is also a small but well-kept and well-shaped yew tunnel arching over each of the two paths that lead to the beautiful little 13th-century parish church, which has a pretty oak-shingled square spire and bell turret.

This churchyard is situated on a small hillock overlooking the surrounding countryside and outside the main village. It sits above a tributary of the River Arun and there is a holy well about 100 m (330 ft) away – both can be reached by following a small track that slopes down to the river and is signposted from the churchyard entrance.

The vast trunk, which is over 7 m (23 ft) in girth, has several openings, allowing you to step inside the cavernous interior

Eastling Yew

27

Location. In the churchyard of St Mary's Church, Kettle Hill Road, Eastling, Kent ME13 0AY. OS grid ref. TQ965565

Rounded and surrounded

Twelve mature yews, some with beautifully fluted trunks, surround this small flint church, which in parts dates from the 11th and 12th centuries. Arranged in two semicircles, they almost completely encircle the building. But that's not all – there is also a truly exceptional yew growing close to and overshadowing the church porch on the western side.

In 1937, Arthur Mee wrote in *The King's England – Kent*: 'all round the church is a ring of yews. If Eastling had nothing else to see we should come to look at the kingly yew near the porch, which must have been a veteran when Chaucer rode to Canterbury.'

From some angles, this yew's huge bulging trunk looks almost round – widest in the middle and tapering above and below this. It is deeply fluted, full of dips and dimples of various shapes and sizes, and at the base, the fissures disappear down into the ground.

The trunk is hollow and contains many internal aerial roots, some quite large. They fill much of the huge void and are visible through a gap on the north-west side, which is itself slowly becoming blocked by them. The main part of the canopy is made up of a huge single central stem, which breaks higher up into many smaller branches, filling the middle of the canopy. Lower down, the branches spread sideways, filling out the lower canopy and balancing the shape of this stunning yew.

You can also see a young sapling grown from a cutting of this ancient yew outside All Saints' Church in Wyke Regis, Dorset. This was planted to celebrate the Queen's 90th birthday in 2016, as part of the church's Living Churchyard Trust project.

The distinctive bulging, bottle shape of the trunk reaches a girth of at least 9 m (30 ft) at its widest point

28

Farringdon Yews

Location. In the churchyard of All Saints' Church, Church Road, Farringdon, Hampshire GU34 3EG. OS grid ref. SU712354

Big brothers

Dating from the 12th century, this church sits in a small, more or less circular churchyard, which is home to two great male yews. When you enter through the Victorian lychgate and follow the path towards the church porch, the first great yew you come to is the larger of the two, standing on the left-hand side of the path, to the south-west of the church tower. This tree is completely fragmented into four large, separate sections, each full of holes. Growing on a slight mound, these sections lean away from the church, around the vast hollow centre. There is a worn path right through the heart of the tree, where many feet have trodden.

Looking up into the tree's crown, you can see an old metal rod and a length of wire that were put in place long ago to keep the sections together, although I'm not sure how reliable they are now. Most of the sections have very little contact with the ground, as there are holes at the base of each of them.

The smaller of the two great yews stands to the south of the church, close to a large Georgian stone cross. This tree is also hollow, but unlike its fragmented brother, the trunk's circumference is almost complete, with just one large opening. It has two substantial limbs that stretch out on each side of its hollow centre, plus a huge central limb, each with many small branches, making a fairly tall tree considering its relatively short trunk.

The smaller yew has an almost intact trunk with just one large vertical split revealing its hollow core

The larger yew has a wonderfully gnarled and craggy trunk, split into four fragments that lean away from the church

Gilbert White, the famous naturalist and ornithologist, writer of *The Natural History and Antiquities of Selborne*, published in 1789 and still in print today, lived in the neighbouring parish of Selborne and was curate of All Saints' Church for more than 20 years. In his journal in 1781, he wrote that the larger yew had a girth of 30 ft (9 m). Nearly 240 years later, it still measures much the same, which is no great surprise due to its fragmented state.

29

Florence Court Yew

Location. Florence Court, Enniskillen, County Fermanagh BT92 1DB.
OS grid ref. NV272002

A true original

This is the 'mother of all Irish yews', the original plant from which all other specimens of Irish yew (*Taxus baccata* 'Fastigiata') in the world are derived. Irish yews differs from common yews in having a very upright growth habit, with denser foliage, and the needles grow all around the twigs, rather than in flat rows.

Growing in such a damp, humid spot near a small river, the tree is covered in moss, not just on its trunks and branches but among its foliage too

In the 18th century, tenant farmer George Willis from nearby Aghatirourke was out hare coursing on land close to his home and came across two young yew plants growing in the wild that looked very different in character to the usual common yew. The accounts of exactly where the trees were found differ slightly, but all agree they were on the slopes of Cuilcagh Mountain above Florence Court, sometime between 1740 and 1760.

George Willis dug up the two saplings he'd found and brought them home in his pockets. One he planted in his own garden. That yew tree lived for around 100 years, eventually dying in about 1865. The other plant he gave to his landlord, William Cole, the 1st Earl of Enniskillen, and it is this tree that survives today, some 250 years later, on the Florence Court estate.

The tree stands in the corner of a grassy area above a small river, surrounded by woodland of predominantly native trees, with a few huge silver firs from old plantings mixed in. It is reached by taking a short, pleasant and well-signposted walk from the visitor centre.

Oddly though, this female tree no longer has the tight, upright appearance you would expect of an Irish yew. Instead it has a fairly open crown and is multi-stemmed. This is thought to be mainly down to the thousands of cuttings that were taken from the tree in the 19th century, as well as the very damp conditions, which have encouraged moss to colonise the tree's branches and trunks, restricting needle growth and in turn leading to thinning of the crown. Still, reports indicate that the tree looked much the same a century ago.

Irish yew trees became widely available from 1818, when they were listed for sale in the catalogue of Hackney nurseryman Conrad Loddiges. So this great old tree is the ancestor of the tens of thousands of Irish yews around the world today, which have all been grown from cuttings taken directly from this tree or its descendants.

Fortingall Yew

30

Location. **In the churchyard of Fortingall Church, Fortingall, Aberfeldy, Perthshire. OS grid ref. PH15 2EX.**

Great Scot

I first visited the famous Fortingall Yew in 2000, as I wanted to see what was reportedly the oldest tree in Europe – said to be anywhere between 3,000 and 9,000 years old!

These original estimates of the yew's age were partly calculated on two measurements taken of the trunk's circumference in 1769. Lawyer and naturalist Daines Barrington recorded the yew's multiple trunks at nearly 16 m (52 ft), whereas naturalist Thomas Pennant, who also visited the tree, noted in his book *A Tour In Scotland* (1769) that: 'In Fortingal church-yard are the remains of a prodigious yew-tree, whose ruins measured fifty-six feet and a half in circumference' – which is 17.5 m. Neither man recorded the height at which they measured the trunk's girth, which may account for the disparity.

The path that leads to the tree has a legend etched into it that reads: 'Up ahead stands Fortingall's oldest resident, a 5,000-year-old yew tree'. However, over the years the estimated age of this tree has been greatly reduced and now sits somewhere between 2,000 and 3,000 years.

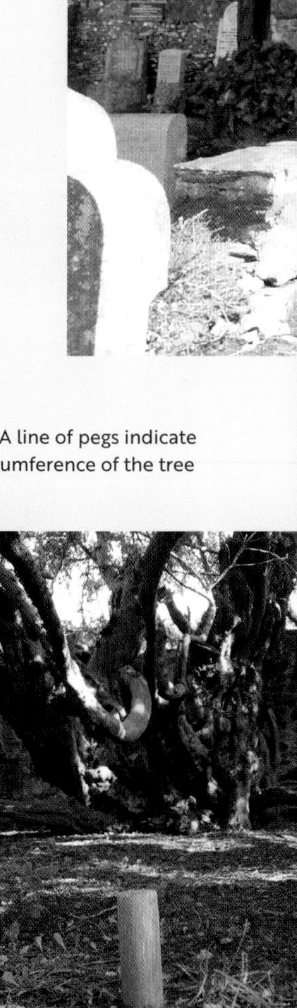

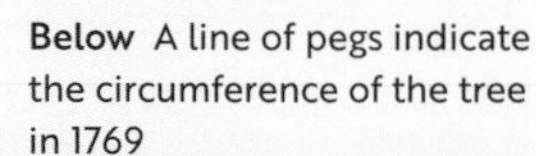

Below A line of pegs indicate the circumference of the tree in 1769

Nevertheless, it is still the most celebrated and well known of the UK's ancient yews.

Sadly, this celebrity among trees has had to be caged for its own protection.

A low enclosure was first built around it in the late 1700s, then replaced with a higher wall that is still standing today. Openings fitted with metal railings provide viewing points. However, this large wall greatly reduces the visitor experience, making the tree quite dark, although without it the yew may not have survived centuries of souvenir hunters keen to cut bits off.

Above The wall was built around the tree to protect it from overzealous souvenir hunters, and it has done its job well, but at the expense of limiting your view of the tree

Despite this, very little now remains of what was once a mighty yew. On one side, parts of the tree are supported on stonework, and on the other it is propping itself up on the enclosure wall, pushing it out. A line of round pegs in the ground shows where the circumference of the trunk was in 1769, when its greatest girth measurement was taken.

The Fortingall Yew is a male tree, but strangely one small branch in the outer part of its crown has become female and has started to produce berries. This is a rare phenomenon, but not unheard of, and has been occasionally recorded on other male yews. The reasons why this happens are not fully understood.

The area that surrounds this venerable tree is also rich in ancient archaeological sites and is the geographical centre of Scotland. In a field close by, about 1 km (0.6 miles) away as the crow flies, are several circles of standing stones, while the Croft Moraig Stone Circle is just a few kilometres to the east. Local legend also claims that Pontius Pilate was born in Fortingall, and as a child played within the shadow of this famous tree!

31

Goostrey Yew

Location. **In the churchyard of St Luke's Church, Goostrey, Cheshire CW4 8PE.**
OS grid ref. SJ779700

Completely stumped

This is a very strange looking tree – really just a living stump with lots of tall, thin new growth. There is a ring of black Victorian rope-top edging stones surrounding its base and, when I visited it, the tree was festooned with lights.

Growing right outside the church doors, this male yew has an absolutely huge, wide trunk that is only around 1.5 m (5 ft) high and made up of mainly dead wood and debris, rotted down into humus. All of the live growth is sprouting from around the perimeter of this huge stump. There is one limb growing out from near the centre, but this is more dead than alive and is defying gravity to remain there at all.

The tree is said to have provided arrows for the Cheshire bowmen who fought in the Battle of Poitiers in 1356. Looking at what's left of it, this is highly possible.

The presence of this ancient yew may be an indication that the mound on which St Luke's Church stands was a focal point for the local community in early history. Stone and bronze axe-heads have been found here, and barrows within the parish boundary suggest that the area has been inhabited as far back as the Iron Age.

The odd shape of this tree is possibly a result of branches being removed for arrow-making in medieval times; yew wood was also widely used for making longbows

32

Gwytherin Yews

Location. In the churchyard of St Winifred's Church, Gwytherin, Conwy LL22 8UU.
OS grid ref. SH876614

Sticks and stones

This churchyard is the focus of a legend dating back to the 7th century and maybe even earlier, although the church that stands here today was built in the 19th century. It is also home to three ancient yews.

The path leading to the church porch runs between two of these giant trees, both females. Their canopies touch above the pathway, creating a sheltered archway above you as you head towards the west end of the church.

The tree on the left is the largest – a fragmented yew of great proportions. It has three main sections and some of its lower branches have been cut back to prevent them reaching the ground and resting on the gravestones below. The internal faces of all the fragments are callusing with new wood, and the tree looks to be in relatively good health.

The yew on the right-hand side of the path has a huge fluted trunk, much of which is hollow, with aerial roots filling the void. Some of the lower limbs are thin, but still very large, twisting out from the main trunk, almost at right angles. Many of the lower shoots on these, as on its partner opposite, have been pruned back to prevent them touching down on the path.

Further along the path, at the rear of the church, is a third female yew, which once would have been a huge tree but is now just two main fragments. One of these is just the dead remains of a once thriving section, showing signs of old fire damage. The living fragment of this tree, although partly covered in ivy, is remarkable healthy, with lots of new growth and a fairly thick canopy.

There is a fourth yew in the churchyard – the smallest and the only male tree. It grows on the high boundary to the north-east of the church, above a small group of houses and a little stream.

There are also four standing stones in the churchyard, on the north side of the church, aligned east to west. One of them, the most westerly, has a Latin inscription carved into it, and is of a style dated to

The beautiful and characteristic deeply fluted trunk is common in many ancient yews

the 5th or 6th century, but the stones may be much older, perhaps from the Bronze Age.

The legend associated with this site tells the story of Gwenfrewi (anglicised as St Winefride), who scorned the advances of Prince Caradoc. He was furious and beheaded her, and her head rolled down the hill, triggering an earthquake, which caused the ground to open up, and water then gushed out. This is said to be the origin of the sacred well at Holywell (in Flintshire) with its miraculous healing waters. Winefride's uncle, St Beuno, then brought her back to life. In the 7th century, she subsequently became the abbess of an early convent here, where she was eventually buried.

The current church, which dates from the 19th century and is now deconsecrated, is thought to stand on the foundations of St Winefride's convent.

Left The largest of the yews has long ago split into three sections, with fresh wood now covering most of what used to be the hollow interior

Below The two remaining fragments of this female yew, thriving after fire damage

Hafod Yews

33

Location. In the churchyard of St Michael and All Angels' Church, Hafod Estate, Pont-rhyd-y-groes, Ceredigion SY25 6DX. **OS grid ref.** SN768736

Green goddesses

The five old yews in this churchyard are of differing ages, and although none would be classed as ancient, they are the oldest yews locally. The two eldest stand behind and to the west of the church. This is a sheltered spot in the middle of mixed woodland on the Hafod Estate, which is mostly owned by Natural Resources Wales. Because of the surrounding woodland and high local rainfall, moisture is retained in this enclosed site, and as a result the trees are green with mosses and ferns. The largest of the yews is divided at a low level, about 1 m (3 ft) from the ground, into two large trunks that grow away from each other, making a large V-shaped forked tree. Verdant, mossy growth almost completely covers the twin trunks and continues far up into the branches.

Mosses and ferns thrive in this damp site, adding a velvety covering to the trunk and branches of the old yews

Both trunks of this female tree are beautifully and deeply fluted. The tree is in very good condition, evenly branched and mainly solid, but has a little hollowing at the base on the left-hand side, when viewed from the church.

The church, dedicated to St Michael and All Angels, is well off the beaten track. It was built in 1620, replacing a nearby earlier chapel of ease dating from the medieval period. It was then rebuilt in 1803 in the Gothic style, and badly damaged by fire in 1932. As part of the restoration, some of the fragments of the beautiful original Flemish stained glass have been fitted into the chancel windows.

There is a small carpark just above the church with interpretation boards showing the Hafod Estate's many picturesque views and experiences, as well as waymarked walking trails, which include a Gentleman's Walk and a Lady's Walk. According to the guide, the Lady's Walk is a little less strenuous

Hambledon Yews

34

Location. In the churchyard of St Peter's Church, Hambledon, Surrey GU8 4DS.
OS grid ref. SU970390

Watch out for the witch

Nestled in the Surrey Hills, St Peter's Church is mainly Victorian in date and sits on high ground above the village. The original church on this site dates from the 11th century, and its churchyard is home to two large yews – both are male and grow on the south side of the church.

The smaller of the two trees is just as impressive as its larger companion. It is a beautifully proportioned tree with a superb large fluted trunk covered in brownish-red flaking bark.

Its wide, solid trunk looks like it's made up of many smaller pieces fused together. The sides of the trunk are more or less parallel for the first 3 m (10 ft), then the main leader gradually begins to taper off towards its top. The crown breaks at this junction and, as is the case with many of the old yews, the lower branches grow towards the ground in an attempt to root, or layer, themselves. The branching throughout the rest of the tree is nicely spaced, giving it a wonderfully statuesque shape overall.

The older and larger of the two trees grows on an area of sloping ground in the south-east corner of the churchyard. It has an exceptionally large and hollow trunk, with a huge cavernous opening that invites you to peer in. Above the opening, the solid part of the trunk continues to rise skyward, from a height of about 2 m (6 ft), then growing up vertically like a giant chimney above an open fireplace. At around the same height, the lowest branches emerge – these limbs are large and thick, growing out from the main trunk at many different angles, the lowest ones dipping down among the gravestones that surround the tree. A lot of new sapwood is being produced around the openings and the tree looks to be in very good health.

Local legend in Hambledon records that the spirit of an old witch lives in the hollow of this tree and that if you walk around the interior three times, she is sure to appear!

Right The vast hollow interior of the older tree is reputedly home to the spirit of a witch

Below The smaller yew has a dramatically fluted trunk that looks like lots of narrow vertical stems fused together

Hampton Court Palace Maze

35

Location. In the grounds of Hampton Court Palace, East Molesey, Surrey KT8 9AU.
OS grid ref. TQ157687

Prepare to be a-mazed

The world-famous yew maze in the grounds of Hampton Court Palace is the UK's (and probably the world's) oldest surviving hedge maze. Even though it was originally planted with hornbeams, which were replaced with yews in the 1960s, it is still a unique and historic 300-year-old feature and is just as much fun to explore as it was when first created. So if you like a challenge and think you have a good sense of direction, it's well worth the moderate price of entry.

The maze was originally laid out in around 1700 and follows the design of George London and Henry Wise, under commission from William III. As you might expect, it has needed many repairs over the past three centuries. It was originally planted out in what was the Wilderness, which was a managed formal area, so not as wild as its name might suggest. The maze you can visit today is the only surviving part

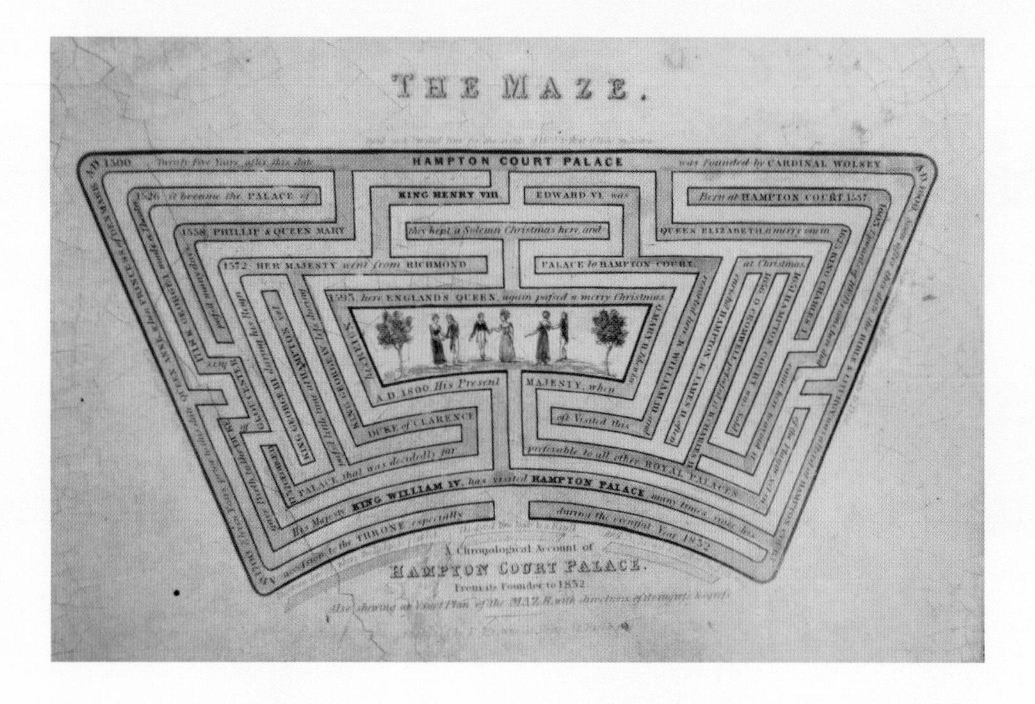

of the original Wilderness created in the 18th century, and was one of two or more mazes within it. In fact, the whole of this area would have seemed almost like one giant maze, with lots of tall hedges separating it into a series of enclosed rooms.

Trapezoid in shape, the maze covers approximately a tenth of a hectare (third of an acre), with around 800 m (half a mile) of paths. The yew hedges are clipped to around 2.5 m (8 ft) tall, and even though they are a bit thin in places, you still get the feeling of being lost – all too easily. On most days, you constantly hear people calling out 'Which way now?' and 'I can't find you!'

The challenging layout is referred to as a puzzle maze or multicursal maze. Unlike a unicursal maze with only one path leading to the centre, this maze has many dead ends. It should take around 20–30 minutes to complete, depending on how many of the dead ends you find!

Above and opposite
Originally laid out three centuries ago, the maze is still as puzzling as ever

Overconfidence can lead you astray of course, as highlighted in the 19th-century humorous novel *Three Men in a Boat*, by Jerome K Jerome:

> *'Harris asked me if I'd ever been in the maze at Hampton Court. He said he went in once to show somebody else the way. He had studied it up in a map, and it was so simple that it seemed foolish – hardly worth the twopence charged for admission. Harris said he thought that map must have been got up as a practical joke, because it wasn't a bit like the real thing, and only misleading. It was a country cousin that Harris took in. He said:*
>
> *"We'll just go in here, so that you can say you've been, but it's very simple. It's absurd to call it a maze. You keep on taking the first turning to the right. We'll just walk round for ten minutes, and then go and get some lunch."'*

The story goes on to say that after walking a good two miles, they realised they were lost and eventually had to call for the old keepers to get them out. Today's visitors will find the maze is still just as tricky, but the entry fee has gone up a little over the past century.

In addition to the yew maze, there are some very old yew trees (thought to be around 300 years of age, like the nearby maze). They are an impressive sight, radiating out behind the palace in the Great Fountain Garden, creating avenues down to the Long Water.

Entry fee applies to the maze and gardens.

Avenues of neatly clipped yew trees – each one a slightly different shape – line the gravel paths of the Great Fountain Garden behind Hampton Court Palace

36

Holeslack Wood Yew

Location. For parking – outside St John's Church, Parkend Lane, Helsington, Cumbria LA8 8AS. OS grid ref. SD488889

Woodland wonder

Dwarfing the nearby trees, both in stature and girth, this mighty old yew can still be hard to find as it's hidden away in the depths of Holeslack Wood

Reaching this well-hidden tree, which is tucked away deep in Holeslack Wood, involves a pleasant and adventurous stroll (see below for details).

It is a male yew, and can be quite an eerie sight when you first come across it, with its huge, bulging, fluted lower trunk, branching from low down. It has three main limbs that are straining up to the light. The central one is the largest and divides again into three. Many of the lower branches are either dead or dying.

The tree is hollow at ground level, with many surface roots snaking out from its base. Although it is the largest standing yew in these woods, there are many other old yews lying on the woodland floor, toppled by the ravages of time.

Around this monumental standing yew are other younger trees, still probably in their hundreds of years old, as if keeping a watchful eye over this patriarch of the woods.

If you visit in late spring, Holeslack Wood is awash with fragrant bluebells (*Hyacinthoides non-scripta*). It is also home to many other native plants that are indicators of ancient woodland, including carpets of dog's mercury (*Mercurialis perennis*), enchanter's nightshade (*Circaea lutetiana*), ferns such as the scaly male fern (*Dryopteris affinis*) and mosses such as the lanky moss (*Rhytidiadelphus loreus*), which covers the stones and boulders on the woodland floor. Holeslack Wood is part of the Sizergh Castle Estate, owned and managed by the National Trust.

Directions To reach the tree, it's an easy walk from the parish church of St John's in Helsington (limited parking opposite the churchyard). Facing the church, walk along the track to your right, keep left at the end, then pass through the gates to Holeslack Farm. Continue down a steep part-concreted path and you will pass the farmhouses on your right. Ahead is another gate, with two semi-derelict medieval barns visible beyond. Pass through this gate and take the path immediately on your right, which goes through a small orchard. Here you will find a small gate leading into the woods.

Once through the gate, turn left and continue down the track for about 50 m (160 ft) to an old fallen female yew on the right, which looks like it has been down for some considerable time, and miraculously is still alive and growing! Walk along the track for another 20 m (65 ft) or so, until you find a young yew on the left of the track. Turn right here onto a small track, which will lead you to the wonderful ancient yew.

Hope Bagot Yew

37

Location. **In the churchyard of St John the Baptist Church, Hope Bagot, Ludlow, Shropshire SY8 3AF. OS grid ref. SO588740**

Hope springs eternal

In springtime, the banks on either side of the small lanes leading to the pretty village of Hope Bagot are carpeted with wildflowers. And as you approach the churchyard, you get your first glimpse of this mighty yew and two other notable yews that grow close by, high up on the bank that marks the boundary.

Clinging valiantly to its steep bank on the churchyard boundary, the Hope Bagot Yew stands above a holy well, whose waters can reputedly heal eye complaints

Entering this ancient oval-shaped churchyard from the south side, through a small gate from the adjacent village hall carpark, you climb some steps and are then greeted by the most beautiful little Norman church, surrounded by wildflowers. Many of its original Norman features are still intact, including the south doorway and an incised round chancel archway inside.

The ancient yew stands on the northernmost edge of the churchyard, above the spring or holy well, which is a grotto-like feature surrounded by stones. Sitting precariously on top of the bank, the tree seems to know it is in a potentially tricky spot, so has sent out large limbs that reach out for more than 15 m (49 ft) on the uphill side and come down to the ground. These will hopefully root themselves, thus anchoring the tree for many more centuries to come.

Viewed from the side facing the church, you can see that this impressive tree is completely hollow, but has many aerial roots growing inside its cavity, down into its decayed base. Like its main roots, these will be reaching down into the spring's nourishing waters below, securing the tree's hold on life.

Inside the hollow trunk are many votive offerings, left by those who have made a pilgrimage to what, for many, is a sacred site and tree. Both are considered symbols of immortality, rebirth, fertility and much more.

Entering the churchyard through the lychgate, there is a sign pointing off to the right, leading you to the holy well, which is directly beneath the ancient yew. The water from the well is said to be 'good for sore eyes'. It is also thought that the well was used for baptisms, which is why the church is dedicated to St John the Baptist.

Iffley Yew

38

Location. In the churchyard of St Mary's Church, 123 Church Way, Iffley, Oxfordshire OX4 4EG. OS grid ref. SP527034

Curiouser and curiouser!

The bark of this old yew is deeply fissured and twisted, and its trunk is completely hollow, with several holes that are partially blocked up with stone slabs and cement. It is these curious openings that are said to have given Lewis Carroll (who was a maths tutor at the University of Oxford's Christ Church College, just a few miles away) the inspiration for *Alice in Wonderland* (published in 1865), which was originally titled *Alice's Adventures Under Ground*. Illustrations of this tree from that period show it with the opening very evident.

In 1857, the yew was pictured in *The Art-Journal* with the accompanying description: 'The church-yard contains an aged yew tree – so aged that no stretch of fancy is required to believe it was planted when the first stone of the sacred structure was laid.' The church dates from the 12th century.

The broad canopy nestles on the ground, due to the tree's unusually short trunk and many low branches

It's easy to imagine the White Rabbit disappearing into the large hole in the base of the hollow, gnarled trunk

This yew stands on the south-east side of the church, between it and an old stone cross with a medieval shaft. It can be difficult to get right up close to the tree, as it has a short trunk with dozens of long, low branches growing out at right angles. The overall crown of the tree is wide and spreading, but it does appear to be suffering some stress, which is evident as areas of die-back at the ends of many of the twigs. Hopefully, as is often the case with old yews, it will bounce back to health again.

The church itself is said to be one of the finest surviving Romanesque parish churches in England, with a high, narrow structure that today looks almost as it did when first built, about 850 years ago. Around the south doorway, you'll find some wonderful, fine carvings of figures of fantasy: centaurs, a merman, a green man and others.

Between the yew tree and the chancel is a 13th-century grave slab thought to be that of a female hermit known as Annora, who reportedly lived for nine years in a hermit's cell here.

39

Itchen Abbas Yew

Location. In the churchyard of St John the Baptist Church, Itchen Abbas, Hampshire SO21 1BJ. OS grid ref. SU534327

Thereby hangs a tale

A once huge male yew sits in the eastern corner of this small churchyard. It's still a big tree, but over the years many of its largest upright limbs have been removed, the bases of which are completely hollow. Still, it has made lots of new growth, which has begun to hide many of these old amputations, and it isn't until you're almost under the tree that they become obvious. Some large lower limbs have also been removed, and the scars of one that would have grown over the low flint boundary wall are very evident.

The short, wide, gnarly trunk has hollows in it, with small aerial roots running down into the cavities. Many small branches grow low down and around much of the trunk. The base of the old main upright stem has bark that is amazingly rippled and looks interwoven, almost like solid latticework. The tree's canopy is noticeably one-sided, with the half facing the lane to the manor and the river being very thin.

The tree's canopy is noticeably thinner on the side facing away from the church

Earth has been removed from the base of the gnarled trunk, which should help to prevent rotting

It is clear that much of the soil that had built up around the base of the trunk has been recently removed, which I'm sure will benefit this old man greatly. Having soil piled up against the trunk increases the likelihood of basal rot.

Close to this old yew lies the gravestone of John Hughes, who according to local legend was one of the last men in England to be hanged for horse theft. He was a local gypsy who was caught stealing a horse and saddle. He was tried and executed in Westminster, but laid to rest here in 1825 by Reverend Robert Wright, who befriended Hughes and promised to bury him under this remarkable tree.

The church itself is Norman in appearance, but was actually completely rebuilt by the Victorians in the 1860s in 12th-century style. This replaced a previous church that probably dated from the late 11th or early 12th century, judging by the pieces of stonework that still survive. However, some records indicate that it may originally have been built as early as 1092.

Kenn Yew

40

Location. In the churchyard of St Andrew's Church, Kenn, Devon EX6 7UR.
OS grid ref. SX921856

Beers all round!

This male yew is truly colossal, with the largest girth of any yew in Devon. Although there is a fragmented yew of almost similar size in nearby Payhembury (see p 164), Kenn's yew looks a lot bigger because it is almost intact at the base.

This massive tree is made up of two distinct sections, which subdivide into eight main limbs growing almost vertically, apart from one that shoots out horizontally on the east side of the tree and is propped up on an old telegraph pole. The central part of the trunk, at just over 1 m (3 ft) in height, has an internal platform that can only be viewed from the east side of the tree, along the main path. This platform is made up of decayed debris from the tree.

The internal parts of the main limbs are covered over with new sapwood, showing that it has obviously been hollow for some time. Much of the rest of the trunk is coated in young growth, dead leaves, ivy and cobwebs. Overall, the tree looks very healthy, with a huge canopy and good leaf coverage on all of its branches.

This tree, like many other ancient yews across the country, stands close to a source of water. The churchyard is on a patch of high ground at the bottom end of the village, close to a stream and water meadows.

Entering the churchyard from a small carpark on the south-east of the church, you pass a beautifully illustrated information board showing the natural history of the churchyard, including the wealth of wildflowers that grow there. Like many old churchyards, this is a managed stronghold and a haven for local flora.

Also in the churchyard, against the east boundary wall, is a small hut, the Bier House. A bier was a cart used to carry coffins into churches for funerals. Incorporated into the south wall of this small building is a stone archway decorated with flowers, said to date from the 1300s and originally to have been the west archway of the church tower.

There is also the other kind of beer to be enjoyed in the 12th-century inn opposite the church!

The Kenn Yew's very wide trunk is still intact at the base – it only becomes hollow from a height of about 1 m (3 ft)

Kingley Vale Yews

Location. Kingley Vale National Nature Reserve, Chichester, Sussex PO18 9BE. OS grid ref. SU824088

Enchanted forest

Kingley Vale is home to one of the largest and finest yew forests in western Europe – at the bottom of the valley is a large grove of around 70 veteran and ancient yews, with more in a second group a little higher up.

It's an amazing experience to walk among all these wonderful trees – you really need to come here in person to fully appreciate this special place. Over time, the trees' branches have become twisted and forked, like many tortured arms reaching out from the trunks. The arching form of these limbs allows you to walk beneath, through and around the trees. Where they touch the ground, many have rooted, producing new trees and stabilising the parent tree. The new trees that rise from the layered branches are known as 'walking trees'.

There is a good mix of male and female trees in the grove, but the seed-bearing female trees' progeny can only establish themselves beyond the main grove, due to the lack of light within – it's too harsh an environment for any yew seedling to thrive. The seed is dispersed to surrounding areas mainly by birds, to spots where they can germinate successfully, thereby expanding the gene pool of this colony. Still, the grove itself is continuing to spread clonally through the 'walking trees' produced from layering branches. It is this process that gives rise to the arches and tunnels made by these limbs, many of which are now dead, having successfully produced the next generation. This is the main way they would have spread, I am sure, in natural wild yew woods.

The trees come in all shapes and sizes. Some are very tall, although you have to be outside the grove to see their tops. From within the grove, the trunks all look very different – most are hollow, some with fluted surfaces, others covered in burrs, and some bulging halfway up as if they have middle-age spread. The trunks can have a girth of over 5 m (16 ft) and are often no longer vertical, with some leaning at almost 45 degrees.

Kingley Vale was made a National Nature Reserve in 1990, and woodland covers much of its southern slopes. It also has SSSI (Site of Special Scientific Interest) status and is managed by Natural England.

Many of the yews have drooping branches that touch the ground, helping to support the tree and often rooting to produce new clonal offspring

Directions The yews are reached from a small car park in Downs Road, West Stoke. There is a short walk (of around 15 minutes) to a visitor centre, with the start of the main grove just a little further on.

Langley Yew

42

Location. In the churchyard of St Mary the Virgin Church, St Mary's Road, Langley, Berkshire SL3 7EL. **OS grid ref.** TQ004795

Breaking free

'The first thing that strikes a stranger on entering the churchyard of Langley, is a magnificent old yew tree, which appears coeval with the ancient church itself', states Edward Jesse 1847, in his book *Favourite Haunts and Rural Studies.* Today, 170 years later, the tree may well be in a more decayed condition, but it's still a striking sight as you walk along the path towards the church.

In total there are ten yews in this churchyard – five to the north of the church and five to the south. But this old male yew, the one you come across first, is by far the largest and oldest. The iron railing, placed around it many years ago for its protection, resembles the tree itself – old, decayed and no longer offering the protection it once did.

With its short trunk split into two large hollow fragments that both show evidence of earlier fire damage, the tree stands on a small mound that is mostly made up of centuries' accumulation of its own decaying material. The larger of the two fragments, the one nearest the church, divides into two large leading limbs at around 1 m (3 ft) above the ground, continuing up into a healthy, much-branched crown. High up in the canopy you can see a series of old rusty metal chains and cables with tensioning screws and bolts. Some have broken away from their fixing points, so no longer serve their purpose, although I'm sure it was these that kept the two fragments upright for many years.

The church itself, the oldest in Langley, dates from 1150, with some of its original features remaining. The tower was rebuilt in brick in the 17th century, replacing the previous flint tower. The churchyard is the largest habitat for wildlife in Langley. The Heritage Trial leaflet points out that under the old yew tree is the grave of John and Alice Guy and their 11 children!

Despite having split in two, this huge tree is still faring better than the warped and broken iron railings that surround it

Leeds Yew (Kent)

43

Location. In the churchyard of St Nicholas's Church, Lower Street, Leeds, Kent ME17 1RL. **OS grid ref.** TQ825533

A bid for freedom

To the right of the church's massive Norman tower, practically rubbing shoulders with it, is an equally massive female yew. The tree has a really healthy crown, which looks from a distance to be almost as tall as the tower itself. It isn't until you get a bit closer that you realise that this dense canopy is supported by just two pieces of a trunk, fragmented long ago. A large iron rod, which links two metal clamps that are each attached to one of the fragments, holds them together. The canopy is fairly evenly distributed between the two fragments, forming a good overall shape.

Growing on a mounded area close to the churchyard entrance, by the south-west corner of the church, the tree is protected behind 2 m (6 ft) high metal railings, although these are starting to come apart. Both of the tree fragments are covered with ivy, growing up to and beyond where the first branches break above the top of the railings' curved spikes.

Despite its protective cage, the yew is trying to do what yews do naturally, and that is to surround itself with its own clones. To this end, branches are escaping over the railings and touching down on the other side, where they will hopefully layer (or root into the soil) and help to support the tree. Whether this is allowed to continue remains to been seen!

According to the church guide, St Nicholas's Church is mentioned in the *Domesday Book* of 1086, which indicates the presence of a Saxon building on this site at that time. Both the yew and the church occupy the front of a large churchyard of 1.6 hectares (4 acres), which is partly left to grow naturally, with meadow grasses and naturalised snowdrops and daffodils in spring.

The tree has a remarkably healthy, well-balanced canopy, bearing in mind it is supported by just two fragments of trunk

Levens Hall Topiary Yews

44

Location. Levens Hall, Kendal, Cumbria LA8 0PD. **OS grid ref.** SD496851

Clipped into shape

In the garden of the Elizabethan Levens Hall is an amazing collection of topiary yews that are more than 300 years old, dating back to when the garden was first laid out at the end of the 17th century. Remarkably, the garden has survived almost unchanged ever since and has been authenticated by *Guinness World Records* as the oldest topiary garden in the world.

There are more than 100 different topiary shapes in the garden, making a great spectacle when you enter. Some of the forms are very abstract, others are more recognisable. Many are also very large. Not all of them are made from yew – some are box (*Buxus*) – but the largest are all styled from yew. The tallest piece of topiary in the garden is a shaped yew known as the Umbrella Tree, with a wide spreading canopy. It's a living dome with a seat encircling its trunk, and many visitors have undoubtedly enjoyed the shelter it gives from the elements.

Not all the yew topiary is made from the normal green yew – a yellow-leaved cultivar is also used to great effect in shapes such as the Golden Pyramids, where it creates a really good colour contrast against a mainly green background.

Trimming all of the garden's topiary and hedging starts in September and goes on for a good few months. Most of this work is carried out during the garden's closed period, from October to early April.

Topiary – the art of shaping plants – has been around since Roman times. Evergreens are typically used, particularly yew. Yew is a versatile plant and lends itself to repeated trimming, enabling it to be crafted into many forms – intricate formal shapes, as well as weird and wonderful stylised forms.

The 17th-century diarist John Evelyn, a friend of Charles II, claimed to be the first person to bring yew into fashion as topiary, and it may have been displayed in his garden at Sayes Court in Deptford. That garden unfortunately no longer exists, but a small part of the original estate is now a park.

In the world's oldest topiary garden, all manner of curiously shaped yews, many dating from original plantings in the 1690s, stand shoulder to shoulder

There is a charge for entry into Levens Hall Garden, which is open several days a week from spring through to autumn.

Linton Yew

45

Location. In the churchyard of St Mary's Church, Linton, Herefordshire HR9 7RX. OS grid ref. SO66022532

Young at heart

There are two ancient yews in this churchyard, both of which have been hollow for many years. The smaller of the two is on the south-eastern side of the church, close to the chancel, and the much larger one – the biggest yew in the county of Herefordshire – is on the northern side, close to the early 13th-century porch.

When you enter through the lychgate, the larger yew is the first of these two trees you come across. It is an absolutely huge female tree with a wide, bulging and cavernous hollowed-out trunk. The yew's remarkable capacity to regenerate itself can clearly be seen in the hollow trunk of this tree. An aerial root put down many years ago has now turned into a large internal stem, growing almost centrally in the cavity, and is itself quite large at more than 2 m (6 ft) in girth. These aerial roots take over some of the feeding of the tree, as the old trunk slowly breaks up, renewing the tree from the inside. It is the reason why the outer girth of the trunk has stayed almost unchanged for many years, as much of the energy and growth has gone into the new inner stem.

This mighty tree has survived a deliberate fire in 1998, which caused severe damage, from which it is still recovering. When I visited almost 20 years later in 2017, its charred interior was still very evident. Remarkably though, the tree is still growing well and looks to be in relatively good health, considering both the fire and its undoubted age.

The smaller of the two yews is male, and still a tree of grand proportions with many large upright limbs. The trunk is gnarled and full of depressions, with an opening on one side exposing its hollow centre.

This tree is a great illustration of how yews can regenerate themselves by producing aerial roots inside their hollow trunk, which eventually transform into vigorous new stems

Llanfihangel Nant Melan Yews

46

Location. In the churchyard of St Michael's Church, Llanfihangel Nant Melan (A44), Presteigne, Powys LD8 2TR. **OS grid ref.** SO180581

Rowdy rooks and a sleeping dragon

In this old oval churchyard, at the foot of Radnor Forest close to Summergil Brook, stand five notable yews, set around the church: one behind it to the east, two on the south side and two on the west side.

One of the larger trees is no more than a living shell, totally hollow with twisting, rippling and fissured bark. Two others have double trunks, one divided at ground level and one just above. There are no trees on the north side of the church, so it is not fully encircled. However, this may once have been a complete circle of yews, as is often the case at pre-Christian sites.

The local rooks have taken over the tops of these trees – dozens of them becoming extremely vocal as you walk beneath, adding to the atmosphere. The lower branches and surrounding ground are covered with fallen twigs from their nest-building, and the trunks are splattered with white droppings.

Although the church looks very old, it was in fact completely rebuilt by the Victorians in 1846, following the demolition of a much earlier building. It was modelled on a church in Herefordshire and built in the Norman style. The chevron arches above the porch doorway and a window at the west end of the church have survived from an earlier building.

According to local legend, the last Welsh dragon lies deep within Radnor Forest. It is said that four churches were built, of which this is one, to surround the forest and make sure the dragon didn't escape. They were all dedicated to St Michael, the conqueror of the dragon. Apparently, the dragon will awaken and go on the rampage if any of the churches are destroyed and the circle broken.

Right A surprising quantity of twigs, from rooks' nest-building, has accumulated over the years around the base of the yews

Below One of the yews is a hollow shell of rippled and pitted bark, yet it continues to thrive undeterred

Llangattock-juxta-Usk Yew

47

Location. **In the churchyard of St Cadoc's Church, The Bryn, Llangattock-juxta-Usk, Monmouthshire NP7 9AP. OS grid ref.** **SO330096**

Many-toed giant

This is definitely a tree of two halves. It is a large male yew, growing on the western edge of St Cadoc's churchyard, close to the bell tower. It sits in a break in the boundary wall and may once have been within it.

Viewed from inside the churchyard, it looks like many other ancient yews. Large, wide and with a hollowed trunk. It is only when you go round to the other side that this tree's remarkable secret aspect is revealed. On the side that faces outwards into the field adjacent to the churchyard, a series of what are probably large aerial roots are exposed. These must have been put down many years ago, lending the tree vital support and hopefully added longevity, with the hollowed shell of the original trunk having rotted away long ago. Seen from this side, the tree has an almost fairytale magic – it looks like it's standing on long-stilted, many-toed legs, and that it might get up and walk away at any moment!

There is a medieval Welsh poem called *Cad Goddeu* (The Battle of the Trees) that tells a traditional story in which the legendary enchanter Gwydion brings to life the trees in the forest so they can fight as his army in a mythological battle. Perhaps this yew might just be one of the survivors...

With the outer trunk having rotted away on this side, the tree resembles some sort of fantastic creature with lots of many-toed legs

Llangernyw Yew

48

Location. In the churchyard of St Digain's Church, Llangernyw, Conwy LL22 8PQ.
OS grid ref. SH875674

Centre of attention

There are claims that this could be the oldest tree in Wales, although I am not convinced. Be that as it may, it is still a truly impressive, fragmented male tree, with four large sections of the trunk surviving.

As soon as you enter the churchyard, you are faced with the huge crown of this yew tree, which stands on the north side of the beautiful little 13th-century church of St Digain. With its white-washed stone walls and distinctive cruciform layout, the church is unusual for this part of north Wales, where they are more usually single or double nave in style.

The tree's once hollow trunk is now so open that you can walk right through the middle, with the remaining parts of the original tree all leaning out from the centre. The sections furthest from the church consist of three large limbs that are all still joined at their base. One of these reaches out at a right angle, propping itself up among the old gravestones, a full 10 m (33 ft) from where it joins the other two sections.

The section closest to the church is the largest and makes up more than 50 per cent of the overall crown. This fragment also has many branches growing towards the ground among the gravestones, all attempting to touch down, root, regrow and stabilise the remains of this magnificent tree.

Nearby stand the remains of two other old fragments, now just dead shells. There is also an area of old dead epicormic growth, now silvered with age, reminding me of artfully crafted coral. Epicormic growth is something you often see on old yews, sprouting from low down on the trunk – it is usually triggered into life when the upper canopy is thinning or the tree is under stress.

Two standing stones, which may have been early grave markers, can also be found in the churchyard, on the south side of the church between the vestry and south transept. The earliest of the two is believed to date from between the 7th and 9th centuries and is

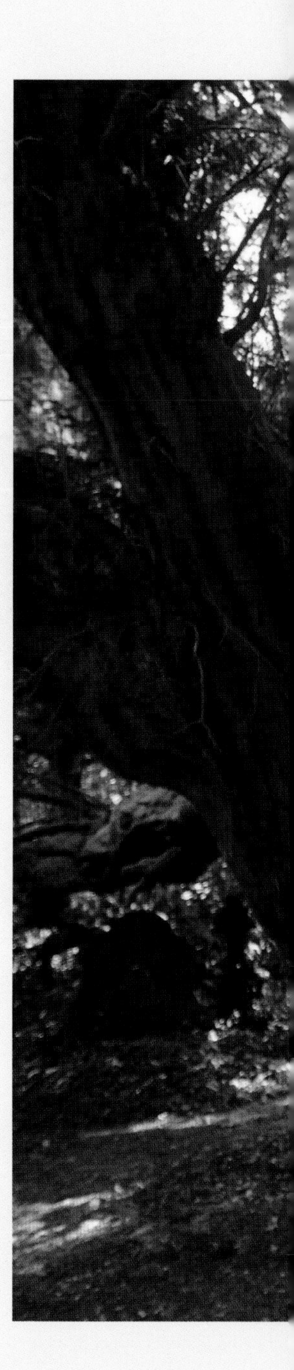

inscribed with an equal-armed cross and punch holes at the end of the cross arms. The second stone is later in date, from between the 9th and 11th centuries – it also has cross arms carved into it, but with a circular centre.

The four remaining live sections of the original trunk are all leaning outwards, creating a very open centre, where, in the mid-1990s, the church's oil tank stood

49

Llansilin Yews

Location. In the churchyard of St Silin's Church, Llansilin, Powys SY10 7QA.
OS grid ref. SJ209282

Circle of life

The church of St Silin's is almost completely encircled by old yews – and was probably once entirely ringed by them, as circles of trees are associated with early religions on sacred sites. Many of the yews in this churchyard are ancient, forming an amazing gathering of the species. There are probably more individual old yews here than in any other churchyard in Wales.

Entering through the metal gates and walking towards the church porch, on your left, north-west of the church, you will see the first of the main yews that encircle the church. This is a male tree – it is hollow, with around a third of its trunk completely open. Much of the cavity is filled with large stones covered in ivy.

Of the four other yews on the north side of the church, three are female and one is male. The male tree is the largest yew on this side. It stands furthest from the church, close to the boundary wall, on slightly raised ground. It has a huge hollow trunk that is well on its way to being fragmented, and in its centre is a large dead section, which would have been part of the original trunk.

Continuing round to the back of the church, to its eastern corner, you find the largest female yew in the churchyard. It is difficult to see where the original tree starts, or indeed finishes! It is a short, stocky tree, with a large central stem, surrounded by smaller fragments, some of which may have been from the original trunk, but they may also be old aerial roots that have developed into stand-alone stems. Apart from one dead branch in its canopy, it appears fairly healthy. Alongside the tree is an old (probably antique) concrete lawn roller, which looks to have been out of use for some time!

The largest male yew here, and probably the oldest, has a vast, fluted trunk that has been measured at more than 8 m (26 ft) in girth

Lockerly Yew

50

Location. **In the churchyard of St John's Church, Lockerley Road, Lockerley, Hampshire SO51 0JJ. OS grid ref. SU298266**

In good shape

Entering the churchyard, you cannot fail to see the impressive large yew tree with its wide, spreading canopy. It dominates the north side of the church and its rectangular churchyard, and stands close to the site of the former Saxon church.

This male tree's short, stocky, deeply fluted trunk shows no obvious outward signs of hollowing. Many of the lower limbs have been removed, which may have been done to raise the canopy above the surrounding gravestones. There are three main upright limbs that break from the main trunk at around 2 m (6 ft) from ground level. These produce most of the tree's canopy. The central limb is hollow at its base where it breaks from the trunk. There are quite a few lower branches that

Standing on a slight mound, the tree has a vast trunk and horizontal branches that reach out over the gravestones from about head height

are smaller and grow out towards some of the gravestones, helping to give the tree its lovely overall shape – tall, wide and spreading, with a fairly healthy canopy.

There is also a second, smaller yew in the north-west corner of the churchyard, alongside the boundary hedge. This tree has a large, elongated hollow opening. Its canopy has lost a large branch, but the gap is already filling with new growth.

The present church, built in 1889 and consecrated in 1890, replaced a smaller Saxon church. There is a model of that earlier building on display in the porch, and in spring its dimensions, between the corner stones that still remain, are marked with daffodils.

51

Long Sutton Yews

Location. **In the churchyard of All Saints' Church, The Street, Long Sutton, Hampshire RG29 1ST. OS grid ref. SU738473**

Village elders

Three fine old yews stand in this churchyard, in the beautiful little village of Long Sutton, with its Tudor buildings and duck pond. It is located adjacent to an ancient trackway known as the Harroway, which runs through the south of England and is one of the country's oldest roads, thought to date back to Neolithic times.

Entering the churchyard you immediately come across the first of the three male yews, on the left of the path, south-west of the church porch. It's a fairly tall tree, at about 15 m (50 ft), with four main upright limbs. The two largest are the most central and almost completely hollow at their base. There is also another huge limb that grows out over the perimeter wall. On the side of the trunk facing away from the church you'll find the main hollow area, where there are lots of aerial roots growing down into the ground.

Up in the canopy is an old metal rod that would have once tied some of the canopy together, but the rod is no longer attached at one end. Nevertheless, its old fixing bracket still encircles the limb like an old rusty shackle on a giant's arm. Some of the higher branches on the church side, above the porch, were sympathetically removed when the church roof was recently retiled.

The second tree can be found behind the church on the north side and is by far the largest of the three yews, with a trunk that has been fragmented for some time. It's in three sections, the smallest of which is completely dead and very unstable. The second-largest section is supported with a wooden prop, taking some of the weight of a limb growing out towards the church. The largest section has three main limbs – the middle one has all but broken away from the tree, but is still attached and rests on the ground. It has many vertical branches growing from it that are almost as tall as the tree's main canopy. The other two limbs of this section are both propped up on wooden posts.

Lying on the ground in this huge tree's hollow centre is an old square metal rod that must have once been a tie in the tree's canopy.

Looking up into the first yew, between the two nearest branches, you can see the metal rod that once helped to hold the canopy together

It probably fell from the decaying wood long ago, as there is still a small piece of dead wood attached to it. This tree has a tremendously broad canopy, which I measured at approximately 18 m (59 ft) wide. At the time of my visit, it also had more artichoke galls on it than on any other yew I have seen. These resemble tiny globe artichokes at the tips of the shoots (see p 13).

The third male yew stands on the south-east side of the church. This tall yew has a much more open canopy with twisting, sinuous branches and is in the process of fragmenting into two sections. The large limb growing towards the churchyard's perimeter wall is propped up to prevent it from breaking away from the hollow trunk. Much of the solid part of the trunk is deeply fluted, and the branches carry many burrs. These are bulging masses of malformed buds and shoots, often seen on the trunk and larger branches of yews – and much prized by wood-turners and furniture-makers for their attractively patterned grain.

Left The largest yew has a broad trunk split into three fragments, and the low branch to the right has almost completely fallen away

Right The trunk of the third yew is gradually splitting in two, and its canopy is very thin

The church itself is believed to date from around the 13th century and has several unusual features. On the south-eastern corner of its outside wall is a mass dial – a plain square area, about 1.5 m (5 ft) above the ground, with a small hole in the centre. This would have held a style, making a shadow to show the time of day. Inside the church is a Norman lead-lined carved stone font. Beyond the altar rails on the south side of the sanctuary is a 12th-century piscina – a small arched opening in the wall with a drain hole, where the priest would pour away the water used to rinse the chalice. And on the opposite wall there is a small square niche called an aumbry, where the chalice was kept.

Lorton Yew (Wordsworth's Yew)

52

Location. Behind Yew Tree Hall, Lorton, Cumbria CA13 9UJ. **OS grid ref.** NY161254

Storm survivor

A poem written by William Wordsworth in 1803, called *Yew Trees*, names two specific yews – the Borrowdale Yews (p 60) and this one in Lorton:

> *There is a yew-tree, pride of Lorton Vale,*
> *Which to this day stands single, in the midst*
> *Of its own darkness, as it stood of yore:*
> *Not loathe to furnish weapons for the Bands*
> *Of Umfraville or Percy ere they marched*
> *To Scotland's heaths; or those that crossed the sea*
> *And drew their sounding bows at Azincour,*
> *Perhaps at earlier Crecy, or Poictiers.*
> *Of vast circumference and gloom profound*
> *This solitary Tree! -a living thing*
> *Produced too slowly ever to decay;*
> *Of form and aspect too magnificent*
> *To be destroyed…*

Known locally as Wordsworth's Yew, the tree as it stands today is only a shadow of what it would have been in Wordsworth's day. During a major storm in the 19th century, its girth was reduced by almost half – and the wood that was brought down was used to make a ceremonial chair for the mayor of nearby Cockermouth.

More recently, in the 1990s, a large limb that used to stretch out over the adjacent stream (Whit Beck) was ripped off during a battering by high winds. From a distance, the damage is not that evident, but looking at the tree side-on you can clearly see it is one-sided. Nevertheless, apart from some dead wood in the top of the canopy, it looks to be in good health. The tree actually stands on a small plot of private land just above the beck, behind what was the site of Jennings Brewery in the 19th century and is now the village hall (Yew Tree Hall). Still, there is a public green space beside it, with an interpretation panel, that provides a good view of it.

As rugged as the Cumbrian fells that surrounds it, the Lorton Yew has stood up to the harsh Lakeland weather for many centuries, and continues to thrive

The tree has several interesting facts in its history. George Fox, the founder of the Quakers, preached under it in 1653. And in the mid-18th century, John Wesley, the founder of Methodism, gave sermons under Lorton's yew when he visited Cumbria.

As far as I am aware, this is the only yew to have a book written solely about it – *Wordsworth and the Famous Lorton Yew Tree* was published in 2004 by a local history society. In 2007, the tree also featured in the BBC series and accompanying book *Meetings with Remarkable Trees* by Thomas Pakenham.

Martindale Yew

53

Location. In the churchyard of the Old Church of St Martin, Martindale, Cumbria CA10 2NF. **OS grid ref.** NY434183

A well-grounded individual

The old female yew tree in the northern corner of the stone-walled churchyard occupies a huge area. From up on the fellside above the church, you can get a good view of the size and spread of this extremely healthy old yew. The main part of the tree comprises two large, almost separate trunks of roughly equal size, joined at their base but unsociably growing away from each other. The individual trunks are both hollowing.

Walking around the back of the church, you step under the canopy of the tree's arching limbs into what feels like a darkened room, made up of its many dipping, twisting branches. The lower branches are grounded on all sides of the tree, but there are two on the most northern side that have reached out, twisting their way down to reach the ground, and filling the entire corner of the churchyard a full 18 m (60 ft) away from the main trunks.

The large tomb that lies below the tree's canopy belongs to Richard Birkett, who was the priest of St Martin's Church during the 17th century and died aged 95 years. He served at the church for 67 years, from 1633 until his death on Christmas Day 1699. His epitaph states that he left the sum of £100 'towards the better maintenance of a godly, sober and religious Minister at Martindale Chapel'.

The first reference to a church on this site is from 1220, although the current building is mainly 16th century, with restoration in 1882 after the roof was damaged by violent storms. It is surrounded by a small rectangular dry-stone-walled churchyard, along a remote narrow and winding dead-end lane in an atmospheric valley on the south side of Ullswater.

Directions To find the church and the yew tree, head for Pooley Bridge on the B5320 and take the unnamed road to Martindale that runs along the eastern side of Ullswater. Pass St Peter's Church, then take the next left to the Old Church of St Martin.

The yew's outstretched arms touch the ground all around, embracing the whole northern corner of the churchyard under its huge shady canopy

Much Marcle Yew

54

Location. St Bartholomew's Church, Monks Meadow, Much Marcle, Ledbury, Herefordshire HR8 2NF. **OS grid ref.** SO657327

Take a seat

This well-known and much-visited yew has a hollow interior fitted with three wooden benches which are said to seat up to 12 people. When I visited it, a blackbird was sitting on her nest in there too! The benches were installed in the 18th century, according to the church wardens' records.

The church building dates from the 13th century, but this ancient yew in its churchyard probably goes back centuries before that date. The tree frames the doorway of the church porch on the south side, and its huge, bulging, deeply fluted trunk is completely hollow, although from the rear of the tree this isn't at all obvious.

In 2006 the tree was given a spruce-up – much of the dead wood was removed and the crown was thinned to reduce its weight, which allowed many of the old metal supports to be taken away. This work has greatly benefited the tree, although some of the lower branches are still held up by old Victorian gas-lamp columns. As with all of these great trees, the many visitors who come to wonder at its size and age unwittingly compact the soil around its roots. So far, though, it appears to be coping.

Inside the church itself you can try to spot the six different carvings of the Green Man, some of which are on the capitals in the nave and date to around the mid-13th century, although the Green Man symbol seems to first appear in British churches in the late 12th century. One of these carved heads has a 'sun wheel' on a chain around its neck, which is an ancient symbol representing the solar calendar, fertility and much more.

Close to the churchyard, to the north, is a small wooded area that hides the earthworks of a motte and bailey castle, first recorded in 1153 as the remains of Mortimer's Castle. It is thought that stones from the castle walls were used to build the church's tower.

Right The bulbous, fluted trunk measures over 9 m (30 ft) in circumference

Below Benches inside the tree's hollow trunk reportedly seat up to 12 people, but that would be quite a squeeze

Newlands Corner Yews

55

Location. Newlands Corner, Drove Road, Albury, Guildford, Surrey GU4 8SE.
OS grid ref. TQ042492

Land of the giants

While researching this book, I have seen some very big and ancient yews of all shapes and sizes, but those at Newlands Corner really are amazing. You can definitely see yews elsewhere with broader trunks, but you will rarely find taller yews or so many large yews in a relatively small area.

These magnificent yews must once have dominated this area, as most of the other types of trees here are much younger. The shape and size of the yews' canopies also suggest that the woodland was once much more open.

Newlands Corner is very popular with walkers as it sits above the North Downs in an Area of Outstanding Natural Beauty (AONB), and offers some tremendous views of the downs to the south from the carpark with its visitor centre and café.

Walking into the woodland behind and to the north of the café, following a marked footpath, among mainly oaks and birches, you will start to see the darker shapes of the yews. But it isn't until you get up to and around them that you really see their magnificence. They truly are a wonder. In these woods the yews have had to grow tall to reach the light, but because they are fairly well spaced, many have made very large, wide canopies too. Considering their overall size these yews are, as a whole, in better condition than any I've ever seen.

The shapes of these wild yews vary widely, but quite a few have what I call the pot-bellied form – narrowest at the base, then expanding quickly to their widest point, and thinning down again towards the point at which the crown starts to branch. They remind me of portly old gentlemen.

These wild yews have a wide range of strange and interesting forms, although there seems to be a tendency towards bulging waistlines

There are more large yews nearby that are worth a visit too. Take a short walk to Merrow Downs (see directions below), where there is a group of very large yews set in woodland and around the edge of a golf course.

A third group can be found at the very end of the carpark, which is part of an ancient trackway called Drove Road, just above the North Downs Way (which at this point is also the Pilgrim's Way).

These three fragmented groups of yews were probably once part of a much larger yew forest on the chalk ridge of the North Downs.

Directions To reach the yews at Merrow Downs, start at the Newlands Corner car park and take the small footpath on the left that leads to the edge of Trodds Lane to the north-east. Cross this and continue along the footpath to Merrow Downs.

Painswick Yews

56

Location. St Mary's Church, New Street, Painswick, Gloucestershire GL6 6QB.
OS grid ref. SO866096

99 and counting

There are 99 (or so) clipped yews in the churchyard of this 15th-century parish church, whose soaring spire and weathercock reach 195 m (639 ft) above sea level and can be seen from afar in every direction. The yews date from the 18th century, so are around 300 years old. They were planted in avenues from the lychgate, through which coffins would be carried on their way into the churchyard, and represent the hope of resurrection. Yew trees have long been associated with rebirth.

This is a beautiful collection of yews that have been clipped regularly for more than a hundred years – there is an archive photograph that shows them in 1902 looking very like they do today. They are all individually different, as you would expect, but some more so than others. Most have clear stems, and their shaped canopies are mainly rounded or conical. Some are probably best described as freeform, while others join to form archways. But as individuals and en masse, they are definitely a spectacle.

The yews are clipped every year to keep them in shape, usually during September. A specialist company with a very big team completes the job in a single day, producing more than 2 tonnes of clippings. This annual tree clipping is not to be confused with a ceremony known as 'clypping the church' (clypping is an Old English word meaning embracing), which also takes place in September, when local children wear flowers in their hair, join hands and encircle the church building.

According to legend, the reason for there being 99 yew trees in the churchyard (although I haven't actually counted them!) is because a hundredth tree would never grow, as the devil would pull it out. However, a young yew was planted on the north side of the church in 2000 to mark the millennium, making it the hundredth tree, and so far it seems to be doing rather well. Although I have it on good authority that there were already more than a hundred trees...

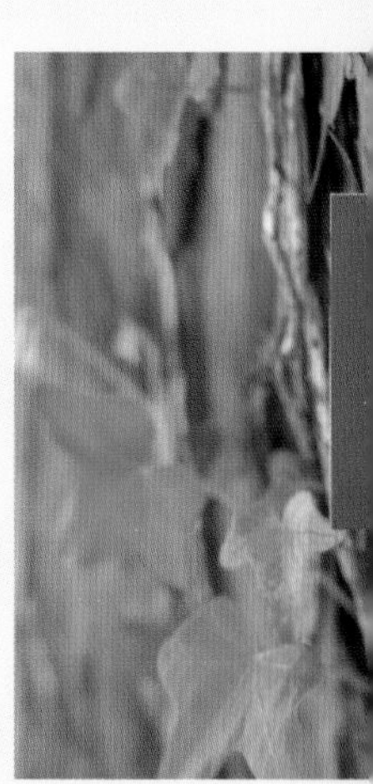

Above and left So how many clipped yews are there in this unique multi-shaped collection? You'll just have to visit and count them for yourself

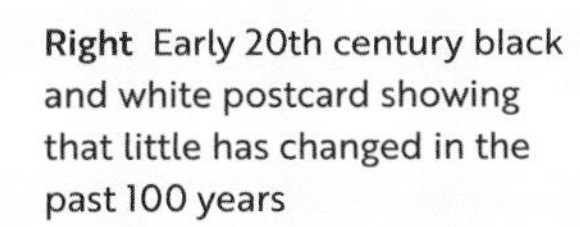

Right Early 20th century black and white postcard showing that little has changed in the past 100 years

Payhembury Yew

57

Location. In the churchyard of St Mary the Virgin Church, Mouse Hole Lane, Payhembury, Devon EX14 3HR. OS grid ref. ST088017

In the path of a lightning bolt

Recognised as one of the 'Great Trees of East Devon', this yew stands in the north-east corner of the churchyard. There is a 19th-century account of its being struck by lightning and split to the ground, into the four sections we see today. These are all of a similar size, with the most northwesterly one being the smallest.

The four fragments of this tree stand on a raised area of ground, all of them growing unsociably away from each other and with their very healthy-looking canopy almost touching the corner of the church. It is obvious that these sections have been apart for quite some time, as they each look almost like separate trees. Their internal sides are calloused over with new bark and they have branches all the way round, although there are still more on the outer sides.

As a result, it was often assumed that the sections were four separate trees growing close together. And reports of at least one section being a different gender to the others seemed to reinforce this misconception, although it is not unknown for parts of a yew to change sex. For example, the Fortingall Yew (see p 104) is a male tree, but has produced berries on one small branch in its crown. However, when I examined this yew's fragmented sections in 2017, I could find no evidence of any female shoots. The sections are all male.

A few metres away from this large fragmented tree is another much smaller yew, which has a short trunk that branches from low down. On one side it is hollowing and a new internal stem is clearly visible. In girth, this tree is of similar size to the largest fragment of the ancient yew. This second tree is a female.

Thought to pre-date the nearby church, the Payhembury Yew has survived a lightning strike that is said to have split its trunk into four

58

Pennant Melangell Yews

Location. In the churchyard of St Melangell's Church, Pennant Melangell, Llangynog, Powys SY10 0HQ. OS grid ref. SJ024265

Princes and pilgrims

Within the dry-stone wall of this circular churchyard are six yews, four of which are notable and very large – two males and two females. The first stands on mounded ground on the right as you enter through the small stone lychgate, south-east of the church. This is the second-largest of the four yews and has a large cavity in its front, as well as the dead remnants of the old branch that would have grown from this point. Close to this tree, so that their canopies are almost touching, is the largest of the yews. This one has a very wide, short trunk and four large, main, upright limbs, while a fifth limb at its front has long-since broken away. This tree is also standing on mounded ground.

On the church's west side is one of the two female yews. It has a deeply fissured, hollow trunk that is large enough for a person to stand inside, and many internal aerial roots have been produced in this cavity. The canopy is a good shape – large and with lots of branching.

The final yew, and the smallest (though not by much), is still a large tree. It stands against the stone boundary wall to the south-west of the church. There are lots of low branches on this tree, some of which are touching the ground, so will probably root, or layer, at some time in the future.

There is a spring and a well on the nearby wooded slope to the north of the church, and a small river on the east side. The ancient stone church, which was rebuilt in the mid-12th century, is sited on some of the flattest ground between the spring and the river, which would have made an ideal spot for an early sacred site. This has been a place of pilgrimage from at least the 7th century, and is thought to have been a Bronze Age site too. You will find it at the end of a long, winding, mostly single-track lane (with few passing places), in a steep-sided valley about two miles from Llangynog.

The village of Pennant Melangell itself takes its name from St Melangell. Legend has it that she was a young Irish woman, the daughter of a Celtic prince, who left Ireland to escape an arranged

Aerial roots can be seen coming down into the hollow centre of the larger female yew

marriage and went to live in a valley in the Berwyn Hills. A hare that was being chased by a local chieftain and his hunting party took refuge under her cloak. After explaining her story to the chieftain and why she came to be there all alone, living as a hermit in the valley, she was gifted the land in the valley and promised to protect it from hunting. Her gravestone and a very rare 12th-century shrine are part of this site, and although the shrine was damaged during the 16th century, it has now been restored. It still draws pilgrims here to this day.

Set on a mound, the second-largest yew has a gaping hole in its trunk where a low branch once grew

Pulpit Yew

59

Location. In the churchyard of St James' Church, Nantglyn, Denbigh, Denbighshire LL16 5PL. **OS grid ref.** SJ004621

Speakers' corner

It is a rather precarious climb, but once you reach the top of the Pulpit Yew's steps you can deliver your speech right across the churchyard

This extraordinary tree, like many other hollow yews, has had its long-ago-hollowed trunk used for a secondary purpose. But somehow this one seems the most fitting.

The ancient Pulpit Yew is the larger of the two significant male yews outside the parish church of St James, in the remote village of Nantglyn. It stands close to the porch on the south side of the church, with the smaller yew close by.

The little polygonal churchyard in which the yews grow was historically even smaller, and was originally rectangular with rounded corners.

As its name suggests, this huge yew has had its hollow bole turned into an outdoor pulpit. The lectern at the top is reached by a set of stone and Welsh slate steps, and there is even a built-in hand rail.

From this pulpit overlooking the churchyard, many open-air sermons must have been preached. Most notably, it is thought to have been used by John Wesley, the famous 18th-century Methodist minister.

The trunk of this remarkable tree is very red and deeply fluted, with tall upright branches growing all around its hollow centre. The largest and tallest branch is on the right-hand side of the steps.

Considering the amount of building material inside the hollow trunk and the huge number of visitors who come to admire (and climb up inside) this tree, it looks to be in remarkably good shape.

The small, mostly Welsh-slate-covered church of St James is very picturesque. It is thought to be medieval in origin, but little survives from this date and it has been restored many times over the past two centuries. An earlier chapel, which pre-dates this church, perhaps from as far back as the 6th century, was dedicated to the Celtic St Mordeyrn, and was sited a few hundred metres to the east.

Rock Walk Yews (Wakehurst)

60

Location. Wakehurst, Ardingly, near Haywards Heath, West Sussex RH17 6TJ.
OS grid ref. TQ341317

Clinging to the edge

The exposed roots of the old yews are a fascinating sight, snaking over and between the sandstone rocks to keep a tight grip as the surrounding soil has eroded away

This is a unique and spectacular display of old yews, with their exposed roots cascading over and straddling large boulder-like rocks along a line of low sandstone cliffs on one side of Bloomers Valley, on the Wakehurst estate.

It is amazing to walk beneath these trees and see the different shapes and forms that their roots take, looking like weird many-tentacled creatures. On a normal yew tree, at best you might see a few surface roots, but here the yews' roots are trying to get a foothold in the soil below the exposed rock, with some of them reaching down and across the bare sandstone for quite a few metres.

It is thought that these yews date from the 19th century, although there is no exact record of planting, and that they were originally growing in sandy soil, which has eroded away over the centuries, leaving their amazing root systems grasping onto the rocks. Many ferns, lichens, liverworts and mosses also grow on and between the sandstone outcrops, all of which add to the magical, slightly eerie atmosphere. After it has rained, the roots take on more colourful hues and this sheltered area often becomes quite damp and misty – this and the changing light enhance the experience of this wonderful place.

Rock Walk is short walk of about 10 minutes that hugs the base of the cliffs, which are made of Ardingly sandstone and reach up to 15 m (50 ft) high in places. This natural geological feature is part of the High Weald, which extends from East Sussex through Wakehurst and on into Surrey. Wakehurst is owned by the National Trust, and for the past 50 years has been managed by the Royal Botanic Gardens, Kew.

There is an entry charge to Wakehurst, but entrance is free for National Trust members (parking charge applies).

61

Seatoller Yew

Location. **On a hillside above the National Trust carpark, off the B5289 Honister Pass, near Seatoller, Cumbria CA12 5XN. OS grid ref. NY245137**

High and mighty

This fine old multi-stemmed upland yew grows in the remnants of an old dry-stone wall, near a footpath. It is close to the ruins of a stone building that may have been used by slate miners in the 17th century.

The tree has two main sections – one side has three major limbs and the other, closest to the stone ruin, has eight. This tree was probably cut back severely many years ago, effectively coppicing it just above ground level. This has given rise to the form it now has, with a wide, spreading canopy made up of numerous stems that fan out from the base.

Growing on this damp hillside, the tree is surrounded by ferns and its stems and branches are clothed in patches of moss, giving the already richly coloured bark added interest, especially when damp. The tree is in pretty good health, apart from a few dying branches, mainly on the opposite side as you approach it.

The tree stands amid beautiful countryside, above the village of Seatoller. From up on this hillside looking south, you can see Great Gable, one of the highest mountains in the lake district, peaking at 899 m (2,949 ft) above sea level. Nearby Johnny Wood, which contains predominantly oaks, is a Site of Special Scientific Interest (SSSI).

Directions Getting to this yew requires a short, mainly uphill walk of around 15 minutes. Park in the National Trust carpark adjacent to Seatoller Farm. Start your walk by turning right out of the carpark, following the road past the farm and cottages until you reach a pair of large slate gateposts on the right. Pass through these and follow the path on through a kissing gate in a dry-stone wall, then keep going along this path, which is signposted 'Coast to Coast', through another kissing gate. Continue on for about 30 m (100 ft), then turn immediately right after the stiled gate and take the lower grass path to the old stone ruin and the upland yew.

The damp climate of the Lake District ensures the tree never goes short of water and is always covered in plenty of verdant moss

Silken Thomas Tree

62

Location. In the grounds of St Patrick's College, Maynooth, Co. Kildare W23 TW77.
OS grid ref. SF945968

A rebellious streak

This is probably the oldest native tree in Ireland. When you enter the college grounds, you'll see two large and imposing yews on the lawns, one each side of the main path leading to the college buildings. The larger one, on the left, is known as the Silken Thomas tree – it is the elder of the two yews and thought to be around 800 years of age.

This female tree has a relatively short trunk, measured at more than 6 m (20 ft) in girth in 2009 by the Tree Register of Ireland. It is deeply fluted and covered in wispy young growth. The trunk divides into two main large limbs, which form a tall, wide, spreading crown that has become quite thin in recent years. The smaller of the two trees has a more upright, solid-looking trunk that is also deeply fluted, with a smaller but denser crown.

In 1534 Thomas Fitzgerald, known as Silken Thomas, raised an army to avenge the death of his father, whom he had been told (wrongly) had been executed in the Tower of London. Renouncing his allegiance to Henry VIII, he rebelled. But the rebellion turned out to be a hopeless struggle, and it is said that he played a lute under the boughs of this yew the night before he surrendered to the king's commander, Lord Leonard Grey, in 1535. Fitzgerald and his five uncles were all executed at Tyburn in 1537 for their offences against the Crown, even though he had been granted a pardon on surrender by Grey.

Fitzgerald, the 10th Earl of Kildare, died at only 24 years of age. He is thought to have been given the name Silken Thomas by his army of warriors, because of his fancy clothes and the silk trappings adorning his horse.

Now standing peacefully on a manicured lawn, this yew was more daring in its younger days, giving shelter to a doomed rebel nearly 500 years ago

63

Staunton Yew

Location. In the churchyard of St James's Church, Ledbury Road (A417), Staunton, Gloucestershire GL19 3QS. **OS grid ref.** SO781292

Split personality

Of the three male yews that grow in the churchyard of the 12th-century St James's Church, the grand old yew closest to the church tower, on the north-west side, is by far the largest. Its massive fragmented trunk is very close to separating into two large sections that are leaning away from each other. Between these, another large section is missing all of its upper branches, these having been removed, probably having died.

Around a third of the way up into the tree's canopy is a thick steel wire that holds the two sections together. Despite the completely hollow trunk, and only being held up by two relatively thin and fragile limbs, the canopy looks pretty healthy with lots of good growth.

The huge trunk, with its deep hollows and flaking bark, seems to twist and lean away from the church low down, but at the top it straightens up and parts have grown back towards the building. However, these have been pruned to avoid damage to either party.

Curiously, there is some old (probably Victorian) metal railed fencing at the side of the tree – it is just one section and appears to be engulfed by the tree. There is also a small brass plaque nailed to the trunk claiming that the tree is over 2,000 years old.

Close by is a beautiful 17th-century timber-framed dovecote, set on the edge of a small duck pond.

With a hollow trunk that's close to splitting in two, this massive male yew still manages to support a healthy canopy

64

Stedham Yew

Location. In the churchyard of St James Church, Mill Lane, Stedham, West Sussex GU29 0PR. **OS grid ref.** SU863225

Grand old man

The guide to the church states that this old yew measures 12 m (40 ft) in circumference, is 2,500 years old, and is one of the oldest in the country. It is certainly a tree of huge proportions and impressive longevity, but not the size and extreme age claimed, although dating yews is always contentious.

A male yew, it stands close to the churchyard entrance and south-west of the church. It has lots of upright limbs growing from all around the outer circumference of the tree, including many lower down. Some of these have developed from aerial roots growing down inside the hollow trunk. The outside of the trunk is covered in lots of fresh green growth.

Below The vast hollow trunk is sprouting fresh green foliage right down to the base

Above The tree's wide, healthy canopy is almost perfectly conical in shape, stretching out unhindered across the churchyard

Looking into the huge central hollow of the tree, through the large slanting opening in the trunk, is like looking into a spiritual space with amazing walls like no others you have ever seen. They are textured with swirling, gnarly patterns in the dead wood, left by centuries of slow decay. There is a small hole, like a boat's porthole, opposite the main opening, enabling you to see right through the tree. Inside the trunk, on both sides of the opening, new wood has been laid down, growing over the dead and decaying wood.

The yew and the church sit on an area of ground higher than the lane below, running along the southern boundary of the churchyard. This must be the older and original part of the church grounds, with the churchyard now extending to the opposite side of the lane as well.

The church is located at the north end of Stedham village and south of the River Rother. The present building, which is largely Victorian, stands on Saxon foundations thought to date from 1040. You can see some of the Saxon masonry at the base of the walls. In the church porch is a stone coffin, complete with stone lid, that was found in the church walls and is thought to be pre-Saxon.

65

Stoke Gabriel Yew (Devon)

Location. In the churchyard of St Mary and St Gabriel, Church Walk, Stoke Gabriel, Totnes, Devon TQ9 6SD. **OS grid ref.** SX848571

Doing what comes naturally

This ancient and unmissable yew spreads its branches from the local-stone boundary wall of the churchyard all the way across to the church itself – a distance of around 30 m (100 ft). What makes this female yew different from most other churchyard yews is that it has been allowed to grow pretty much naturally, which is normally only seen in wild, woodland yews. It has had minimal pruning of its branches – just enough to keep it clear of the main path that passes beneath the tree's vast canopy.

The trunk looks completely solid and sits on a raised area that has been surrounded by a low stone wall. It has two main central limbs that are absolutely huge – they have fused together a short way up, before dividing again. They continue up and form into the magnificent tall canopy, almost eclipsing the height of the church tower.

The best way to get under the yew's canopy is from the north church porch and out through the back of the churchyard, where the canopy looks like a long wooden arched tunnel, dark but inviting.

Many of the yew's limbs are propped, and the largest has a steel cable bracing it, keeping those passing beneath it safe, and the tree intact. Some of the propped limbs twist and turn to form very interesting shapes.

I have not seen another churchyard yew that has been allowed the freedom to do what comes naturally, spreading and layering its branches around themselves. Many have rooted in between the churchyard's tombs and grave stones.

There are two wooden plaques at the base of the tree. One shows a timeline that names some of the historical events that have occurred in the tree's lifetime, and the other a traditional rhyme claiming that the tree will grant you your wishes. It reads: 'Walk ye backward round about me, seven times round for all to see; Stumble not and then for certain, one true wish will come to thee'.

According to the church guide, in the 11th century there is a reference to this yew already being several hundred years old. I think it is certainly likely to be over 1,000 years of age.

Propped branches allow you to pass beneath the yew's vast canopy

The church of St Mary and St Gabriel in Stoke Gabriel was listed in the *Domesday Book*, and sits high above and overlooking a mill creek of the River Dart. The church tower, as with many of the old churches, is the dominant structure and dates from the 13th century. It rises to a height of 20 m (65 ft). The yew that dominates this churchyard matches the Norman tower in height.

Stow-on-the-Wold Yews

66

Location. In the churchyard of St Edward's Church, Church Street, Stow-on-the-Wold, Gloucestershire GL54 1BB. OS grid ref. SP190257

On sentry duty

Author J. R. R. Tolkien is said to have got his inspiration for the gateway to Moria in *The Lord of the Rings* from seeing these two yews flanking the door to St Edward's Church. Whether or not this is true, the trees do have a very magical, other-worldly feel, and once you've seen illustrations of the entrance to Moria, the association does seem quite plausible.

There is very little factual information about this pair of yews, which appear to guard the arched doorway into the small north porch of this Cotswold parish church. There are 13th-century mouldings hiding behind the trees and a Victorian lamp hanging between them, for added atmosphere. Although the left-hand tree's buttress roots (which look like a giant lion's paw) are bigger, both trees have a trunk girth that is almost identical, within 2.5 cm (1 in) of each other.

I have read that these trees are thought to be the last in an avenue of yews that once ran up to the doorway in the 17th or 18th century, perhaps similar to those at St Mary's Church in Painswick (see p 162). However, I can find no real evidence for this. It is certainly a bit of a mystery why the trees were planted so close to the entrance. Considering how tight up against the stonework they are, it is fortunate that little or no damage appears to have been caused, mainly because the buttress roots are growing forward, away from the church walls.

The trees, which are both male, will hopefully flourish for many more years to come, without damaging either the church or themselves. They are a great tourist attraction and even on the cold and snowy January day when I visited, I wasn't the only one there to admire them.

These two guardian yews stand so close to the church porch that it's amazing they haven't damaged the foundations or stonework

Strata Florida Yew

67

Location. In the churchyard of St Mary's Church, Strata Florida, Pontrhydfendigaid, Ceredigion SY25 6ES. **OS grid ref.** SN746656

Immortalised in flowery verse

The most famous of all the medieval Welsh poets, Dafydd ap Gwilym, is reputed to be buried under the large female yew that stands beside St Mary's Church. He is said to have lived and died in the adjacent Strata Florida Abbey, a former Cistercian monastery, in the 14th century. Strata Florida translates from Latin as Vale of Flowers.

The ancient yew can be found to the north of St Mary's Church and has a low stone wall around its base. Most of the central part of its trunk is hollow, with some smaller areas of strong, healthy growth behind the cavity. At the base of the tree is a stone marking it as the grave of Dafydd ap Gwilym, although this is disputed. There is also a claim that he was buried 48 km (30 miles) away at Talley Abbey.

Gruffudd Gryg, one of Gwilym's contemporaries, wrote a poem in the 14th century that gives some possible validity to the yew tree burial site:

Yr ywen i oreuwas,
Ger mur Ystrad Fflur a'i phlas;
Da Duw wrthyd, gwynfyd gwŷdd,
Dy dyfu yn dŷ Dafydd

This yew-tree for the best of men,
Near the walls of Strata Florida and its halls;
God's blessing on you, happy tree,
For growing as a house for Dafydd.

Historically, there are said to have been many more yews growing here, so it is also possible that he was buried on this site, but under another tree, now long gone.

The adjacent Strata Florida Abbey is a ruin, having been dissolved in 1539 under Henry VIII's suppression of the monasteries. Today the site is cared for and managed by the Welsh government's heritage agency Cadw, and there is a real air of tranquillity in this valley. An even earlier abbey, dating from the late 12th century, once stood two miles from here, on the banks of the small river Afon Fflur.

This tranquil 'vale of flowers' seems like a fitting place to bury Wales's finest medieval poet, Dafydd ap Gwilym, beneath the mighty old yew

As you walk around the ruins today, you can still get a good impression of how breathtaking the building was, with a huge, grandly decorated, carved entrance arch leading into what are now the low walls outlining the original layout. You can see the bases of large structural columns, chapels with elaborately decorated tiled floors, and the gravestones of monks and Welsh princes.

In the north-west corner of the abbey, in the north transept, is a large slate memorial stone to Dafydd ap Gwilym.

Admission to Strata Florida Abbey is free from November to March, with a small charge from April to October.

68

Tandridge Yew

Location. In the churchyard of St Peter's Church, Tandridge Lane, Tandridge, Surrey RH8 9NJ. OS grid ref. TQ374511

A deep-rooted mystery

When you see this tree for the first time, it is definitely a wow moment. The wide bole is fluted from its base, where at around 1.5 m (5ft) it divides into three huge, individual fluted trunks that are unusually tall for a tree of this age – certainly over 1,000 years old.

Tantalisingly, there were old reports that during archaeological work around the church some old Saxon foundations were discovered, with a deliberate stone vaulted section built over the tree's roots to avoid damaging them. If true, this would potentially date the yew to pre-Saxon times. However, when Tim Hills of the Ancient Yew Group asked the Surrey Archaeological Society to investigate, they could find no archaeological or historical evidence to support this earlier claim.

When I first saw this tree, I was unsure whether it was a group of trees that had grown together or a fragmented tree that had grown back together. But I am pretty sure now that it was either coppiced (cut to the base) at some time in its very distant past or that it had suffered some early damage, as the three main limbs are all of a similar size, so probably similar in age. But whatever the ancient cause, this has become a truly stunning tree.

Today this impressive multi-stemmed giant, with its hollowing trunk, is hidden away in a churchyard on a hill, but this is surrounded by a motorway and busy roads. How different the area must have looked in the tree's youth – at best with a few trackways and probably a lot more woodland.

Even if it doesn't date from pre-Saxon times, this vast and awe-inspiring yew is still at least 1,000 years old

Tisbury Yew

69

Location. In the churchyard of St John's Church, Church Street, Tisbury, Wiltshire SP3 6NH. OS grid ref. ST943291

Heart of stone

As you approach this tree, the first thing that strikes you about it, even before its size and possible age, is the vast amount of concrete that fills its hollow trunk. There must be literally tonnes of concrete in there, reaching a height of more than 2.5 m (8 ft). The render on the concrete has clearly been patched up over the years and there are signs of a pinker render beneath the current surface. I wonder if this was used to blend in better with the colour of the bark.

Before the concrete was added in the middle of the last century, Sir Thomas Dick Lauder, in 1834, wrote that there was enough room in the hollow of this trunk for 17 men to breakfast inside, and that it was entered through a rustic gate. Having visited the tree, I think 17 would have been a very cosy breakfast!

Above the concrete-filled trunk are two main limbs. The largest of these, the one closest to the church, is completely hollow in the lower part and is supported by a wire connected to the other more solid-looking limb, with some of the lower branches cut back to keep them above the ground. Around the base of the trunk is an octagon-shaped line of stones, probably the remains of the low retaining wall that can be seen in old photographs.

Before the concrete was added, aerial roots were apparently developing in the trunk's hollow centre, which could have helped to revitalise the tree. Sadly, these would probably have been destroyed. Nevertheless, the tree looks to be in remarkably good condition.

This huge female tree is not the only yew in the churchyard, but it is the only outstanding one. To find it, enter the churchyard though a small metal gate from Church Street, and head along the path that leads directly to the church's north porch. Once you pass the two female fastigiate yews that flank the path, the ancient yew is just a few metres in on the right.

There has been a religious building on this site – on high ground on the north side of the small River Nadder – since the 7th century.

A monastery was destroyed by the Danes in the 9th century and was probably followed by a Saxon church, then a Norman one, with many parts still remaining in the church that stands here today. This ancient yew has probably witnessed most, if not all, of these changes through the centuries.

Despite the tonnes of concrete filling its hollow trunk, the tree appears to be thriving, with a dense canopy of healthy foliage

70

Totteridge Yew

Location. In the churchyard of St Andrew's Church, Totteridge, London N20 8PR.
OS grid ref. TQ246941

A Roman relic?

This ancient female yew is the oldest tree – and the oldest living thing – in London. It is located on the outskirts of the capital, close to the A1 and A10, which were both originally Roman roads, and this tree is said to date from that era. Its surroundings would certainly have looked very different in those days.

The girth of the trunk was recorded as long ago as 1677, when it measured a little over 8 m (26 ft), and this figure has changed very little over the past 300 years. This is probably due to the many aerial roots growing down inside the hollow interior – the tree's energy is going into these and the canopy, rather than into the old trunk.

Much of the original trunk is now mostly a dead shell covered in old burrs and ivy. All of the growth comes from the hollow interior and is made up of half a dozen or so main limbs that have a very good coverage of foliage. A photograph from the early 20th century shows the yew in very poor shape, so it is much recovered since then.

The tree stands on the west side of the church, close to the wood-clad tower. An old report from 1722 records that a baby boy was abandoned under the shelter of this ancient yew. He was taken in by the church and brought up by the parish, and subsequently apprenticed out at the age of nine or ten.

It is also thought that between the 12th and 17th ceturies a primitive court system, called the hundred courts, which administered the law prior to the county courts that we have today, possibly met under this tree to make their decisions on legal matters.

Thought to have started life during the Roman occupation of Britain (43–410 AD), this gnarled yew is the oldest living thing in London

Ulcombe Yews

71

Location. In the churchyard of All Saints' Church, Ulcombe Hill, Ulcombe, Kent ME17 1DN.
OS grid ref. TQ846497

Bulging waistlines

Four very characterful yews grow in this well-maintained churchyard, which has Norman origins. The largest and probably the oldest is a male tree, growing close to the south side of the church's tower and doorway. It has an absolutely huge, bulbous trunk that looks quite solid, but in fact has a large hollow area. The girth has been measured at around 10 m (33 ft), making it among the largest of any British yews. Most of the central parts of this magnificent tree have long gone and the main growth now comes from smaller upright limbs around the outer part of the tree, masking its tired core. Much of the wispy growth has been trimmed lower down to keep the access clear along the path. At the base of the tree is a brass plaque, set in a stone, stating that this yew was here before the birth of Christ.

Right Dappled sunlight filters through the thin canopy of the huge female yew, highlighting its beautiful rich red bark

Below The vast, bulbous trunk, which measures about 10 m (33 ft) in circumference, makes this one of Britain's largest yews

The second largest yew is a female with a deeper red bark overall and a similar but slightly smaller girth. This is the first yew you come to on the path as you enter the churchyard. It has a huge low, fluted trunk with a hollow that has some small aerial roots growing down inside. Large branches break from low down, some growing out at right angles, but most heading directly upwards.

Further into the churchyard, to the north-west of the church, are two further yews standing side by side, like a pair of rotting pillars. They are of a similar size and both completely hollow, and mostly just shells of dead sapwood. The tree growing closest to the perimeter of the churchyard is a male with a slightly larger girth than the female tree that grows closer to the path. Inside its hollow trunk you can see a large internal stem.

The 12th-century church is of Norman origin, but has been renovated and added to over the centuries. In the south aisle are some medieval paintings, one depicting St Michael and the Devil.

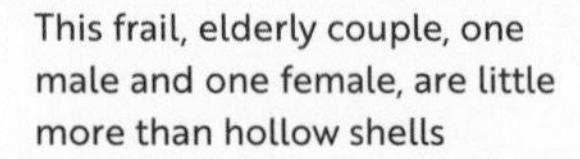

This frail, elderly couple, one male and one female, are little more than hollow shells

Watton-on-Stone Yew

72

Location. St Andrew and St Mary Church, Church Lane, Watton-at-Stone, Hertfordshire SG14 2RJ. **OS grid ref.** TL302188

Stress management

This large and bulging venerable male yew has a deeply fluted trunk that appears to be mostly solid. But around its base you can clearly see some of the hollow interior, through two or more openings, made more obvious by the burrowing of the churchyard rabbits.

The shape of the tree's canopy is wide and spreading, with a wonderful domed top, which is somewhat surprising when you get under the canopy and see its limited limbs. The trunk is relatively short and goblet-shaped, thinnest at its base, then expanding quickly to its bulging, widest point.

Much of the trunk is quite difficult to see, as it's entirely covered in a mass of twiggy growth. Although some is dead, plenty is still healthily growing and green. Most of this would have been produced during times of stress, of which there have probably been many, judging by the interior of the canopy. This stress would have been caused not just by its advancing years, but also by the loss of many of the large limbs that would once have furnished the canopy. Three main limbs remain, surrounded by lots of smaller, lower side-branches around the tree, which together form a very large and healthy-looking exterior, whose dark green foliage shows very little sign of die-back.

The yew can be found to the north of the 15th-century church, which is built on the foundations of an earlier building. It stands on slightly higher ground than the main part of the churchyard with the gravestones. This may have been due to the position of the earlier church, which was perhaps surrounded by a churchyard that was smaller than the present one.

Standing on higher ground overlooking the main part of the churchyard, this yew has a particularly wide and beautifully domed canopy

Waverley Abbey Yew

73

Location. Waverley Abbey, Waverley Lane, Farnham, Surrey GU9 8EP.
OS grid ref. SU868453

Off the wall

This yew is female, with a very short trunk that has a mass of exposed and tangled roots. These sprawl over and across the ruins of part of the old abbey building, looking almost like a long flow of lava. I can only imagine that a seed was inadvertently deposited by a bird who had been feeding on yew berries in the woods that would have surrounded the abbey. This yew seed managed to germinate in the ruins of the wall and slowly grew into the amazing tree we see today. The spot where it is growing would have been the south-east corner of the main building in this 12th-century abbey.

The tree's main trunk starts at around 1 m (3 ft) above the surrounding ground level. It arises directly from the stonework, and is quite short, only around 1 m (3 ft) tall. It then breaks into a multi-stemmed, almost fan-like form with the largest limb being almost central and growing straight up. It has a huge domed canopy that is more than 20 m (65 ft) across and the overall tree looks to be in very good condition.

Waverley Abbey was founded in 1128 by William Gifford, Bishop of Winchester. It was the very first Cistercian monastery in Britain, and was originally home to 12 monks and an abbot from Aumône Abbey in France. In its heyday, Waverley was a thriving and substantial abbey, with a chapter house and monks' dormitory, and around 200 monks and lay brothers living and worshiping within its walls. The whole complex covered around 24 hectares (50 acres) and was bordered on two sides (south and east) by the River Wey. There are earthwork remains of a brewhouse and fish ponds on the site, which was in use for more than 400 years.

In the mid-16th century, the monks of the abbey were dispersed by Henry VIII, during the dissolution of the monasteries. Much of the abbey was dismantled and the stone re-used as building material elsewhere, yet there are mid-18th century drawings showing substantial ruins still standing. So this tree cannot be as old as it looks, as it is growing out of ruins that are no more than 300 to 500 years old!

A yew's location may give a clue to its age – this one at Waverley Abbey stands on the rubble of a wall destroyed less than half a millennium ago, so is at most 500 years old

74

West Kingsdown Yew

Location. In the churchyard of St Edmund King & Martyr Church, Fawkham Road, West Kingsdown, Sevenoaks, Kent TN15 6AY. **OS grid ref.** TQ579633

Set in concrete

Like many ancient and veteran yews, this tree has a hollow trunk, but unfortunately the interior space has been partially filled with concrete. This looks to have been done on more than one occasion, with older concrete visible deeper in the hollow.

Concrete was commonly used to fill tree cavities from the early 1900s through to the 1960s, as it was thought that it would strengthen the tree and prevent disease getting into the hollow. We now know this is not the case – it stops the tree flexing naturally and causes more problems than it cures. Also, in the case of yews, it prevents aerial roots growing down into the hollow centre, where they would normally root into the ground to feed, support and extend the life of the tree.

This is a very upright male yew with a wide, spreading trunk that tapers into the main canopy. The bulging lower trunk has quite a lot of dead twiggy growth on one side and divides into two main limbs at around 3 m (10 ft) up. Both of these have well-balanced branching with plenty of healthy thick foliage, forming a rounded crown. The tree stands on the west side of the church, close to the boundary fence.

The churchyard also contains the remnants of what was once another very old and large yew – with a trunk more than 2m (6 ft) wide – which is sadly now just a dead stump, with two of its clones on either side. You'll find these just in from the lychgate, on the right-hand side of the path.

The church itself is locally known as the 'church in the woods', as it's surrounded on all sides by woodland, unsurprisingly called Church Wood. It is a beautiful little church, built of local stone from the North Downs. It is thought to have started life as a small chapel built for the Saxon lord of the manor, and its oldest parts date back to before the Norman Conquest in 1066.

Below The stump may be long dead, but the two clones that have grown on either side are thriving

Right Filled with concrete, the trunk can't flex naturally and aerial roots can't grow down into the hollow centre to help sustain the tree

West Tisted Yew

75

Location. **In the churchyard of St Mary Magdalene's Church, West Tisted, Alresford, Hampshire SO24 0HJ. OS grid ref. SU650292**

Tall, dark and handsome

This is a colossal tree. Its height, and the girth of its trunk, make it one of the largest yews I have ever seen. The main part of the trunk is hollow, but inside there is an aerial root that has grown spiralling down at an angle and entered the ground – it now forms a sturdy stem with a diameter of more than 90 cm (3 ft).

You won't find many yews that are larger or more impressive than this towering titan, with its huge columnar trunk topped with a vast spreading canopy

The trunk's beautifully fluted exterior is clear of branches for almost the first 7 m (23 ft), which looks incredibly impressive, and the rippling folds give it a muscular appearance. The overall height of the tree is close to 18 m (60 ft).

This remarkable male yew stands on the south side of the church, hidden away in the very tranquil, overgrown churchyard. As you approach the tree along the meandering gravel path, the sheer scale of it starts to become more apparent. You can also see the largest of three cavities around its base, where the trunk is at its widest. It is through this opening into the hollow interior that the large aerial root is most visible. The trunk then reduces in width at around 3 m (10 ft) from the ground, towering up from there. From the other side, the trunk appears to be much more solid and more even in its width, hiding the hollows.

The trunk divides into the tree's two main branches, which are both hollow at their base, and they continue vertically upwards to the tree's current full height. They then terminate abruptly, having clearly lost a good bit of their original length. The tree's crown is large and in a healthy condition.

St Mary Magdalene's is a single-cell Norman church that has been modified and added to over the centuries, on a site that has been used for religious purposes since Saxon times.

Wilmington Yew

76

Location. **In the churchyard of St Mary and St Peter's Church, Wilmington, Polegate, East Sussex BN26 5SW. OS grid ref. TQ543043**

It's a hold-up

Over the years there has been much debate over whether these two old gnarled trunks are in fact two separate trees or a single tree that became hollow over many years and separated into two parts. I think they are one magnificent tree.

There are nine wooden props, mainly old telegraph poles, supporting this female yew. You can also see some huge old chains in the crown, which would once have supported the limbs and held them together. However, they are now no longer fully taut, partly due to the wooden props taking some of the tree's weight, although these would probably not actually hold it completely, if the tree ever decides to move!

The two trunks are now growing and leaning away from each other. The one closest to the 15th-century church porch is the larger of the two and leans at a much lower angle. The combined, spreading canopy of the two trunks casts a vast shadow over the churchyard.

At the base of the tree, on the east side, is a stone, now partly enveloped by the tree. The stone is said to be Roman and was found at the bottom of the old vicarage well by the local village well-digger. It now lies over his grave.

This village is also home to Wilmington Priory, which was built in the 11th century by a French abbot. Although previously in a ruinous state, much of it has been repaired and rooms are now let out to paying guests, with some parts open to the public a few weekends each year.

At the end of the lane from the church, on the outskirts of the village, there is a walk that takes you up to the Wilmington Giant or Long Man of Wilmington. Carved into the downland hillside and measuring over 70 m (235 ft) tall, this vast figure with a staff in each of his outstretched arms really makes a statement. The figure has been renovated many times over the centuries and is now outlined with concrete blocks that are regularly painted to keep him visible. As with many of the ancient yews, the true age of the Wilmington Giant cannot be confirmed, although it is thought to date from the Celtic period.

Several of the nine wooden props supporting this spectacular yew have sunk deep into the bark

INDEX

Images are indicated by page numbers in **bold.**